W. H. Miller

A Tract on Crystallography

Designed for the Use of Students in the University

W. H. Miller

A Tract on Crystallography
Designed for the Use of Students in the University

ISBN/EAN: 9783741163722

Manufactured in Europe, USA, Canada, Australia, Japa

Cover: Foto ©berggeist007 / pixelio.de

Manufactured and distributed by brebook publishing software
(www.brebook.com)

W. H. Miller

A Tract on Crystallography

A TRACT

ON

CRYSTALLOGRAPHY

DESIGNED FOR THE USE OF STUDENTS
IN THE UNIVERSITY.

BY

W. H. MILLER, M.A. For. Sec. R.S., F.G.S.,

FOREIGN MEMBER OF THE ROYAL SOCIETY OF GÖTTINGEN, CORRESPONDING MEMBER OF THE ROYAL ACADEMIES OF TURIN, BERLIN AND MUNICH, MEMBER OF THE IMPERIAL MINERALOGICAL SOCIETY OF ST PETERSBURG, HONORARY MEMBER OF THE SOCIETY FOR PROMOTING NATURAL KNOWLEDGE IN FREIBURG,

AND PROFESSOR OF MINERALOGY IN THE UNIVERSITY OF CAMBRIDGE

CAMBRIDGE:
DEIGHTON, BELL AND CO.
LONDON: BELL AND DALDY.
1863.

INTRODUCTION.

THE following Tract contains an investigation of the general geometrical properties of the systems of planes by which crystals are bounded, and of the formulæ for calculating their dihedral angles, indices and elements, given without demonstration in the last edition of Phillips' *Mineralogy*, or of equivalent expressions in a more convenient shape. To these have been added some theorems which appeared in the *Philosophical Magazine* for 1857, 1858, and 1859. The last two chapters contain concise investigations of the general properties of crystalline forms by the methods of ordinary and of analytical Geometry. These were suggested by a remarkable paper entitled *Sulla legge di connessione delle forme cristalline di una stessa sostanza*, by the Commendatore Quintino Sella (*Nuovo Cimento*, Vol. IV.). The Tract, therefore, besides containing all the theorems of Mathematical Crystallography usually required in calculating the angles of crystals, their elements, and the symbols of their faces, will form, it is hoped, a useful supplement to the Mineralogy, and also to the Crystallography published by the author in 1839. The reader is referred to either of these works for examples, and for an account of the method of using Wollaston's Goniometer.

The angle made by two faces of a crystal will be measured by the angle between normals to the two faces, drawn towards them, from a point within the crystal. The reasons for adhering to this measure of a dihedral angle were given in the *Philosophical Magazine* for May, 1860. It is needless to offer any reasons for retaining the notation, in addition to the remarks made by the late Professor Grailich in his *Krystallographisch-optische Untersuchungen*, p. 6.

The names used in the Mineralogy to designate two of the hemihedral forms of the Prismatic System, and the hemihedral form of the Oblique System, appeared to be inappropriate, and have, consequently, been changed.

CONTENTS.

CHAPTER I.

CHAPTER V.

CHAPTER VI.

CHAPTER VII.

CHAPTER VIII.

CHAPTER IX.

CHAPTER X.

CRYSTALLOGRAPHY.

CHAPTER I.

PROPERTIES OF A SYSTEM OF PLANES.

1. LET OX, OY, OZ be any three straight lines not all in one plane, passing through a given point O; a, b, c any three straight lines given in magnitude; h, k, l any three integers, either positive or negative or zero, one at least being finite.

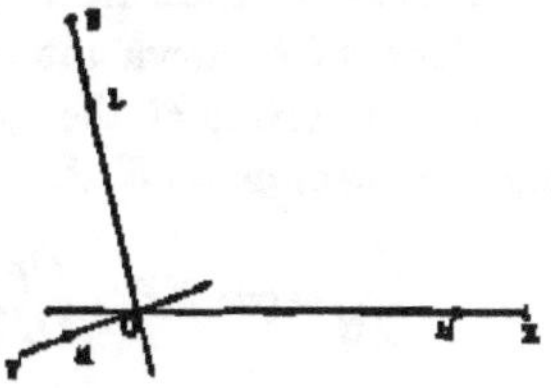

Let a plane HKL meet the straight lines OX, OY, OZ respectively, in the points H, K, L, such that

$$h\frac{OH}{a} = k\frac{OK}{b} = l\frac{OL}{c},$$

OH, OK, OL being measured along OX, OY, OZ or in the opposite directions, according as the corresponding numbers h, k, l are positive or negative. Suppose a system of planes to be constructed by giving to h, k, l different numerical values, the absolute distances of the planes from O being perfectly arbitrary. Let the point O be called the *origin* of the system of planes; the straight lines OX, OY, OZ its *axes*; a, b, c, or any three straight lines in the same ratio, its *parameters*; h, k, l, or any three integers in the same ratio, and having the same signs, the *indices* of the plane HKL; and let this plane be denoted by the symbol $h\,k\,l$. When a numerical index is negative, or a literal index is taken negatively, the negative sign will usually be placed over the index.

It is evident that when one of the indices of a plane becomes 0, the point in which the plane meets the corresponding axis will be indefinitely distant from the origin, and the plane will be parallel to that axis; also, that when two of the indices become 0, the plane will be parallel to the two corresponding axes.

2. Let the axes meet the surface of a sphere described round O as a centre in X, Y, Z; and let OP be a normal to the plane $h\,k\,l$, drawn towards it from O, meeting the plane in p, and the surface of the sphere in P. Then, if the plane $h\,k\,l$ meet the axes in H, K, L,

$$\frac{Op}{OH} = \cos XP, \quad \frac{Op}{OK} = \cos YP, \quad \frac{Op}{OL} = \cos ZP.$$

But $$h\frac{OH}{a} = k\frac{OK}{b} = l\frac{OL}{c}.$$ Therefore

$$\frac{a}{h}\cos XP = \frac{b}{k}\cos YP = \frac{c}{l}\cos ZP.$$

When h is positive, OH is measured along OX, and XOP is less than a right angle; therefore XP is less than a quadrant. When h is negative, OH is measured in the opposite direction, and XOp is greater than a right angle; therefore XP is greater than a quadrant. In like manner YP is less or greater than a quadrant, according as k is positive or negative; and ZP is less or greater

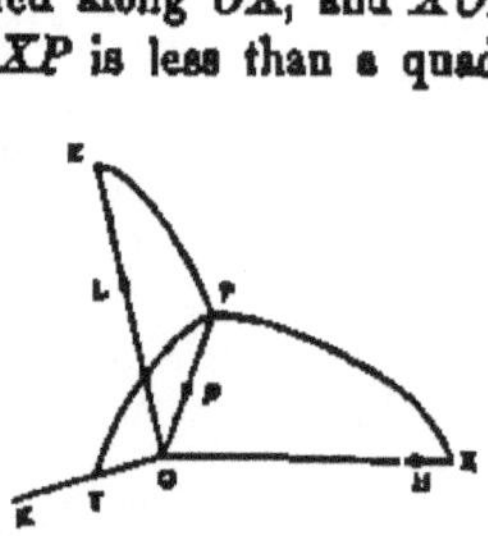

than a quadrant, according as l is positive or negative. The sphere to the surface of which the planes are referred will be called the *sphere of projection*. The outer extremity of a radius of the sphere, normal to any plane, will be called the *pole* of that plane. A plane and its pole will be denoted by the same symbol. The points in which the axes meet the surface of the sphere of projection will be invariably denoted by X, Y, Z.

3. Let A, B, C be the poles $1\,0\,0$, $0\,1\,0$, $0\,0\,1$ respectively; P the pole $h\,k\,l$. Then (2)

$$\frac{a}{1}\cos XA = \frac{b}{0}\cos YA = \frac{c}{0}\cos ZA.$$

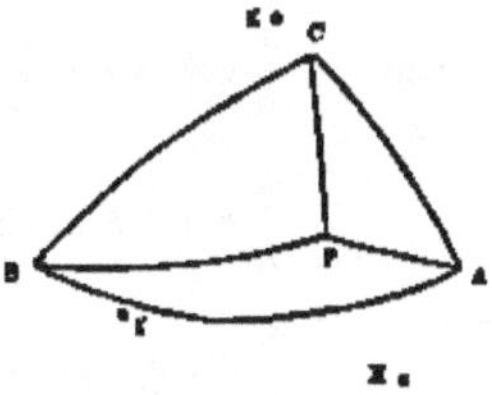

Therefore YA, ZA are quadrants. In like manner it appears that ZB, XB, XC, YC are quadrants. Also (2) since the symbols of A, B, C contain no negative indices, XA, YB, ZC are less than quadrants. Hence X, Y, Z are the poles of the great circles BC, CA, AB adjacent to A, B, C respectively; and A, B, C are the poles of the great circles YZ, ZX, XY adjacent to X, Y, Z respectively. Then, since h, k, l are positive or negative according as XP, YP, ZP are less or greater than quadrants, h will be positive or negative according as P and A are on the same side or on opposite sides of the great circle BC, k positive or negative according as P and B are on the same side or on opposite sides of CA, and l positive or negative according as P and C are on the same side or on opposite sides of AB.

When P is in one of the great circles forming the triangle ABC, the cosine of the arc joining P and the pole of the great circle will be 0, and therefore the corresponding index will be 0.

If a diameter PP' be drawn

$$\cos XP' = -\cos XP, \quad \cos YP' = -\cos YP, \quad \cos ZP' = -\cos ZP.$$

The ratios of the indices of P' will therefore be the same as those of P, but with contrary signs, because P, P' are on opposite sides of the great circles forming the triangle ABC.

4. Since X, Y, Z are the poles of the great circles BC, CA, AB, the arcs XP, YP, ZP are the complements of arcs which divide each of the triangles BPC, CPA, APB into two right-angled triangles. Therefore

$$\cos XP = \sin CP \sin BCP = \sin BP \sin CBP,$$

$$\cos YP = \sin AP \sin CAP = \sin CP \sin ACP,$$

$$\cos ZP = \sin BP \sin ABP = \sin AP \sin BAP.$$

But $\frac{a}{h} \cos XP = \frac{b}{k} \cos YP = \frac{c}{l} \cos ZP$. Hence

$$\frac{a}{h} \sin CP \sin BCP = \frac{a}{h} \sin BP \sin CBP$$

$$= \frac{b}{k} \sin AP \sin CAP = \frac{b}{k} \sin CP \sin ACP$$

$$= \frac{c}{l} \sin BP \sin ABP = \frac{c}{l} \sin AP \sin BAP.$$

From these equations we obtain

$$\frac{k}{b} \sin BAP = \frac{l}{c} \sin CAP,$$

$$\frac{l}{c} \sin CBP = \frac{h}{a} \sin ABP,$$

$$\frac{h}{a} \sin ACP = \frac{k}{b} \sin BCP.$$

5. Let P, R be the poles $h\,k\,l$, $p\,q\,r$; Q any point in the great circle PR, the arcs PQ, PR being measured in the same direction from P.

The spherical triangles PQX, RQX give

$$\cos XP = \cos XQ \cos PQ + \sin XQ \sin PQ \cos PQX,$$

$$\cos XR = \cos XQ \cos RQ + \sin XQ \sin RQ \cos RQX.$$

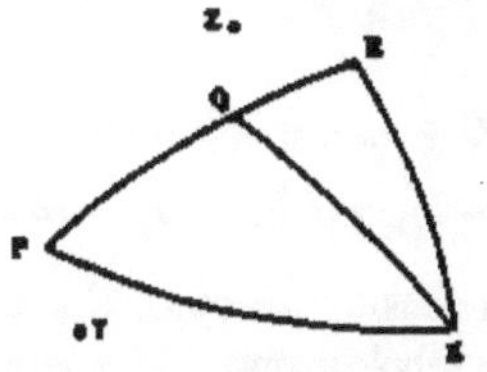

Multiply both sides of the first equation by $\sin RQ$, both sides of the second by $\sin PQ$, and add, observing that when PQ is less than PR

$$\cos PQX + \cos RQX = 0,$$

and $$\sin RQ \cos PQ + \cos RQ \sin PQ = \sin PR.$$

The resulting equation is

$$\cos XP \sin RQ + \cos XR \sin PQ = \cos XQ \sin PR.$$

When PQ is greater than PR, we must interchange Q and R in the preceding equation, which then becomes

$$-\cos XP \sin RQ + \cos XR \sin PQ = \cos XQ \sin PR.$$

Writing $\sin(PR - PQ)$ for $\sin RQ$, in order to reduce the two cases to one, and then substituting Y and Z successively for X, we obtain the following equations:

$$\cos XP \sin(PR - PQ) + \cos XR \sin PQ = \cos XQ \sin PR,$$

$$\cos YP \sin(PR - PQ) + \cos YR \sin PQ = \cos YQ \sin PR,$$

$$\cos ZP \sin(PR - PQ) + \cos ZR \sin PQ = \cos ZQ \sin PR.$$

Whence, by elimination,

$$\begin{aligned}&(\cos YP \cos ZR - \cos ZP \cos YR) \cos XQ\\ &+ (\cos ZP \cos XR - \cos XP \cos ZR) \cos YQ\\ &+ (\cos XP \cos YR - \cos YP \cos XR) \cos ZQ = 0.\end{aligned}$$

But $$\frac{a}{h}\cos XP = \frac{b}{k}\cos YP = \frac{c}{l}\cos ZP,$$

and $$\frac{a}{p}\cos XR = \frac{b}{q}\cos YR = \frac{c}{r}\cos ZR.$$

Therefore,
$$\text{u}a\cos XQ + \text{v}b\cos YQ + \text{w}c\cos ZQ = 0,$$

where $$\text{u} = kr - lq,\ \text{v} = lp - hr,\ \text{w} = hq - kp.$$

The great circle passing through the poles $h\,k\,l$, $p\,q\,r$ may be denoted by the symbol u v w. The numbers u, v, w will be called the indices of the great circle PR. Any three integers in the same ratio as u, v, w, satisfy the equation between cos XQ, cos YQ, cos ZQ, when substituted for u, v, w, and therefore may be used as the symbol of the great circle PR. When u, v, w have a common measure it will be convenient to employ as indices the lowest integers in the required ratio. In cases where there is reason to apprehend that the great circle u v w may be mistaken for a plane or a pole, it may be distinguished from the latter by the symbol [u v w].

6. When three or more planes of the system of planes have their poles in the same great circle, they are said to form a *zone*. The great circle passing through the poles of any two planes not parallel to each other, and which, therefore, passes through the pole of any other plane in the same zone with them, will be called a *zone-circle*. The diameter which joins the poles of the zone-circle will be called the *axis* of the zone. A zone, its zone-circle, and any line parallel to its axis, will be denoted by the same symbol. Hence, the intersections of the planes of a zone, being obviously parallel to its axis, and to one another, may be denoted by the symbol of the zone.

The symbol of the zone containing the planes 0 1 0, 0 0 1, or of a line parallel to the axis OX, is 1 0 0; that of the zone containing the planes 0 0 1, 1 0 0, or of a line parallel to the axis OY, is 0 1 0; and that of the zone containing the planes 1 0 0, 0 1 0, or of a line parallel to the axis OZ, is 0 0 1.

7. Let h k l, p q r be the symbols of any two zone-circles intersecting in the points Q, Q'. Then (5), since Q is a point in each of the zone-circles,

$$\text{h}a \cos XQ + \text{k}b \cos YQ + \text{l}c \cos ZQ = 0,$$
$$\text{p}a \cos XQ + \text{q}b \cos YQ + \text{r}c \cos ZQ = 0.$$

Hence, putting $u = \text{kr} - \text{lq}$, $v = \text{lp} - \text{hr}$, $w = \text{hq} - \text{kp}$,

$$\frac{a}{u} \cos XQ = \frac{b}{v} \cos YQ = \frac{c}{w} \cos ZQ.$$

The indices of each of the zone-circles are integers, therefore u, v, w are integers. Hence, Q, Q' are poles of planes belonging to the system of planes, and common to the zones h k l, p q r.

The points Q, Q' are the opposite extremities of a diameter of the sphere, therefore (3) the indices of Q being u, v, w, the indices of Q' will be $-u$, $-v$, $-w$.

8. It appears that when $u\ v\ w$ is the symbol of the pole in which the zone-circles h k l, p q r intersect, the expressions for u, v, w, in terms of h, k, l, p, q, r, are precisely the same as the expressions for u, v, w, in terms of h, k, l, p, q, r, where u v w is the symbol of the zone-circle passing through the poles $h\ k\ l$, $p\ q\ r$. If the symbols $h\ k\ l$, $p\ q\ r$ be written twice, as below, one under the other, and the letter X three times in the middle three intervals, it will be seen that each of the indices u, v, w is the product of the indices joined by the thick stroke of the corresponding letter X, minus the product of the indices joined by the thin stroke,

$$\begin{matrix} h & k & l & h & k & l \\ & \text{X} & \text{X} & \text{X} & & \\ p & q & r & p & q & r \end{matrix}$$

$$\text{u} = kr - lq, \quad \text{v} = lp - hr, \quad \text{w} = hq - kp.$$

It will sometimes be found convenient to use the symbol $h\ k\ l$, $p\ q\ r$ to denote either the zone-circle containing the poles

$h\,k\,l$, $p\,q\,r$, or one of the poles in which the zone-circles $h\,k\,l$, $p\,q\,r$ intersect, the two cases being distinguished, when requisite, as in (5).

9. Let $u\,v\,w$ be the symbol of the pole Q in the zone-circle p q r. Then (2), (5),

$$\frac{a}{u}\cos XQ = \frac{b}{v}\cos YQ = \frac{c}{w}\cos ZQ,$$

and $$\text{p}a\cos XQ + \text{q}b\cos YQ + \text{r}c\cos ZQ = 0.$$ Hence

$$\text{p}u + \text{q}v + \text{r}w = 0.$$

This equation expresses the relation between the indices of a zone and those of any one of its planes. Any positive or negative integers, including one or two zeros, which satisfy this equation, when substituted for u, v, w, are the indices of a plane in the zone p q r; and any positive or negative integers, including one or two zeros, which satisfy the same equation, when substituted for p, q, r, are the indices of a zone containing the plane $u\,v\,w$.

10. When the zone-circle p q r passes through the pole $u\,v\,w$, we have, by (9), $\text{p}u + \text{q}v + \text{r}w = 0$. Hence, in order to find the poles which lie in a given zone-circle, or the zone-circles passing through a given pole, we must discover the integral values, in which one or two zeros may be included, of x, y, z which satisfy the equation $ax + by + cz = 0$, where a, b, c are the indices of the given zone-circle in the former case, and of the given pole in the latter, not necessarily arranged in the order in which they stand arranged in the symbol. Let the coefficients c, b be prime to each other. Transform $c : b$ into a continued fraction, and let $e : d$ be the last but one of the resulting converging fractions. Then by the solution of an indeterminate equation of the first degree, $y = \pm(eax - mc)$, $z = \pm(mb - dax)$, where the upper or lower sign is to be taken, according as cd is greater or less than be. The value of x being assumed, the cor-

responding values of y and z may be obtained by substituting different positive or negative integers for m.

11. Let P, Q, R, S be four poles in one zone-circle, PQ, PR, PS being all measured in the same direction from P; e f g, p q r the symbols of any two zone-circles KP, KR passing through P, R respectively, neither of which coincides with PR; $h\ k\ l$, $u\ v\ w$ the symbols of Q, S respectively. Then (5)

$$\cos XP \sin (PR - PQ) + \cos XR \sin PQ = \cos XQ \sin PR,$$

$$\cos YP \sin (PR - PQ) + \cos YR \sin PQ = \cos YQ \sin PR,$$

$$\cos ZP \sin (PR - PQ) + \cos ZR \sin PQ = \cos ZQ \sin PR.$$

Multiply both sides of the first, second, third of the preceding equations by ea, fb, gc respectively, and add, observing that P is a pole in the zone-circle e f g, and therefore (5),

$$\mathrm{e}a \cos XP + \mathrm{f}b \cos YP + \mathrm{g}c \cos ZP = 0.$$

Next, multiply by pa, qb, rc respectively, and add, observing that R is a pole in the zone-circle p q r, and therefore

$$\mathrm{p}a \cos XR + \mathrm{q}b \cos YR + \mathrm{r}c \cos ZR = 0.$$

The equations thus obtained are

$$(\mathrm{e}a \cos XR + \mathrm{f}b \cos YR + \mathrm{g}c \cos ZR) \sin PQ$$
$$= (\mathrm{e}a \cos XQ + \mathrm{f}b \cos YQ + \mathrm{g}c \cos ZQ) \sin PR,$$
$$(\mathrm{p}a \cos XP + \mathrm{q}b \cos YP + \mathrm{r}c \cos ZP) \sin (PR - PQ)$$
$$= (\mathrm{p}a \cos XQ + \mathrm{q}b \cos YQ + \mathrm{r}c \cos ZQ) \sin PR.$$

By the substitution of S for Q in the preceding equations, we have

$$(\mathrm{e}a \cos XR + \mathrm{f}b \cos YR + \mathrm{g}c \cos ZR) \sin PS$$
$$= (\mathrm{e}a \cos XS + \mathrm{f}b \cos YS + \mathrm{g}c \cos ZS) \sin PR,$$
$$(\mathrm{p}a \cos XP + \mathrm{q}b \cos YP + \mathrm{r}c \cos ZP) \sin (PR - PS)$$
$$= (\mathrm{p}a \cos XS + \mathrm{q}b \cos YS + \mathrm{r}c \cos ZS) \sin PR.$$

But Q, S are the poles $h\,k\,l$, $u\,v\,w$ respectively, therefore (2),

$$\frac{a}{h}\cos XQ = \frac{b}{k}\cos YQ = \frac{c}{l}\cos ZQ,$$

$$\frac{a}{u}\cos XS = \frac{b}{v}\cos YS = \frac{c}{w}\cos ZS.$$

Hence

$$\frac{\sin PQ}{\sin PS}\,\frac{\sin (PR-PS)}{\sin (PR-PQ)} = \frac{\mathrm{e}h+\mathrm{f}k+\mathrm{g}l}{\mathrm{e}u+\mathrm{f}v+\mathrm{g}w}\,\frac{\mathrm{p}u+\mathrm{q}v+\mathrm{r}w}{\mathrm{p}h+\mathrm{q}k+\mathrm{r}l}.$$

12. It is easily seen that the left-hand side of the preceding equation is positive, except when one only of the zone-circles KP, KR passes between Q and S; or that the arcs PQ, PS, RQ, RS must be considered positive or negative according as they are measured in the directions PR or RP. If we attend to this rule the equation may be written

$$\frac{\sin PQ}{\sin PS}\,\frac{\sin RS}{\sin RQ} = \frac{\mathrm{e}h+\mathrm{f}k+\mathrm{g}l}{\mathrm{e}u+\mathrm{f}v+\mathrm{g}w}\,\frac{\mathrm{p}u+\mathrm{q}v+\mathrm{r}w}{\mathrm{p}h+\mathrm{q}k+\mathrm{r}l},$$

in which the correspondence between the poles P, Q, R, S on the left-hand side of the equation, and the symbols e f g, $h\,k\,l$, p q r, $u\,v\,w$ on the right-hand side, is more easily perceived than in the original form of the equation.

13. $\sin (PR-PQ) = \sin PR \sin PQ (\cot PQ - \cot PR)$,

$\sin (PR-PS) = \sin PR \sin PS (\cot PS - \cot PR)$.

Therefore (11),

$$\frac{\cot PS - \cot PR}{\cot PQ - \cot PR} = \frac{\mathrm{e}h+\mathrm{f}k+\mathrm{g}l}{\mathrm{e}u+\mathrm{f}v+\mathrm{g}w}\,\frac{\mathrm{p}u+\mathrm{q}v+\mathrm{r}w}{\mathrm{p}h+\mathrm{q}k+\mathrm{r}l}.$$

From which, having given the symbols of the zone-circles through the poles P, R, the symbols of the poles Q, S, and the arcs PR, PQ, the arc PS may be found.

14. Putting

$$\tan\theta = \frac{eh+fk+gl}{eu+fv+gw}\,\frac{pu+qv+rw}{ph+qk+rl}\,\frac{\sin(PR-PQ)}{\sin PQ},$$

we have
$$\frac{\sin(PR-PS)}{\sin PS} = \tan\theta.$$

Whence
$$\frac{\sin PS - \sin(PR-PS)}{\sin PS + \sin(PR-PS)} = \frac{1-\tan\theta}{1+\tan\theta}.$$

But
$$\frac{\sin PS - \sin(PR-PS)}{\sin PS + \sin(PR-PS)} = \frac{\tan(PS-\frac{1}{2}PR)}{\tan\frac{1}{2}PR},$$

and
$$\frac{1-\tan\theta}{1+\tan\theta} = \tan(\tfrac{1}{4}\pi-\theta).$$

Therefore $\tan(PS-\frac{1}{2}PR) = \tan\frac{1}{2}PR\tan(\frac{1}{4}\pi-\theta)$.

Whence, having given the symbols of the zone-circles through P, R, the symbols of Q, S, and the arcs PR, PQ, the arc PS may be found.

15. Let m n o be the symbol of the zone-circle PR. Then from (11) and (9) we have

$$\frac{pu+qv+rw}{eu+fv+gw} = \frac{ph+qk+rl}{eh+fk+gl}\,\frac{\sin PQ}{\sin PS}\,\frac{\sin(PR-PS)}{\sin(PR-PQ)},$$

and $mu+nv+ow=0$, two equations from which, having given the arcs PR, PQ, PS, and the symbols of P, Q, R, the ratios of u, v, w, the indices of S, may be found.

16. Let KP, KQ, KR, KS be four zone-circles passing through the pole K; e f g, p q r the symbols of KP, KR; $h\,k\,l$, $u\,v\,w$ the symbols of the poles Q, S in the zone-circles KQ, KS. Let the zone-circle QS meet KP in P, and KR in R. Then

$\sin KP \sin PKQ = \sin PQ \sin KQP$,

$\sin KR \sin RKQ = \sin RQ \sin KQR$,

$\sin KP \sin PKS = \sin PS \sin KSP$,

$\sin KR \sin RKS = \sin RS \sin KSR$.

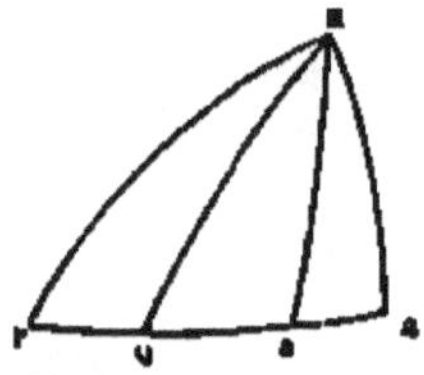

Hence, observing that $\sin KQP = \sin KQR$, and $\sin KSP = \sin KSR$, we obtain

$$\frac{\sin PKQ}{\sin PKS}\,\frac{\sin RKS}{\sin RKQ} = \frac{\sin PQ}{\sin PS}\,\frac{\sin RS}{\sin RQ}.$$ Therefore (12)

$$\frac{\sin PKQ}{\sin PKS}\,\frac{\sin RKS}{\sin RKQ} = \frac{eh+fk+gl}{eu+fv+gw}\,\frac{pu+qv+rw}{ph+qk+rl}.$$

As in (12) the left-hand side of the preceding equation is positive, except when one only of the zone-circles KP, KR passes between Q and S.

17. It may be proved exactly in the same manner as in (13), that

$$\frac{\cot PKS - \cot PKR}{\cot PKQ - \cot PKR} = \frac{eh+fk+gl}{eu+fv+gw}\,\frac{pu+qv+rw}{ph+qk+rl}.$$

Hence, having given the symbols of KP, KR, Q, S, and the angles PKR, PKQ, the angle PKS may be found.

18. Putting

$$\tan\theta = \frac{eh+fk+gl}{eu+fv+gw}\,\frac{pu+qv+rw}{ph+qk+rl}\,\frac{\sin(PKR-PKQ)}{\sin PKQ},$$

we obtain exactly as in (14)

$$\tan(PKS - \tfrac{1}{2}PKR) = \tan\tfrac{1}{2}PKR \tan(\tfrac{1}{4}\pi - \theta).$$

Whence, knowing the symbols of KP, KR, Q, S, and the angles PKR, PKQ, the angle PKS may be found.

19. The symbols of the zone-circles KP, KR being e f g, p q r, and the symbols of the poles Q, S being $h\,k\,l$, $u\,v\,w$, it is sometimes convenient to denote the expression

$$\frac{eh+fk+gl}{eu+fv+gw}\,\frac{pu+qv+rw}{ph+qk+rl}$$

by [e f g, $h\,k\,l$. p q r, $u\,v\,w$], or by $KP,Q.KR,S$, either of which suggests the formation of its numerator. The reciprocal of the same expression may be denoted by [e f g, $u\,v\,w$. p q r, $h\,k\,l$], or by $KP,S.KR,Q$, either of which suggests the formation of its denominator.

20. Let $\dfrac{eu+fv+gw}{eh+fk+gl}\,\dfrac{ph+qk+rl}{pu+qv+rw}=i$. Then (11), supposing PS greater than PR,

$$\sin PS \sin (PQ-PR)=i \sin PQ \sin (PS-PR).$$

But $2\sin PS \sin (PQ-PR)$

$$=\cos (PS-PQ+PR)-\cos (PS+PQ-PR)$$
$$=\cos (2PR-PQ+RS)-\cos (PQ+RS),$$

And $2\sin PQ \sin (PS-PR)$

$$=\cos (PQ-PS+PR)-\cos (PQ+PS-PR)$$
$$=\cos (PQ-RS)-\cos (PQ+RS).$$

Therefore

$$\cos (2PR-PQ+RS)=(1-i)\cos (PQ+RS)+i\cos (PQ-RS).$$

Whence, having given the symbols of KP, KR, Q, S, and the arcs PQ, RS, the arc PR may be found.

In one of the most frequent applications of the preceding equation, PQ is a quadrant, and the equation becomes

$$\sin (2PR+RS)=(2i-1)\sin RS.$$

21. Let *EF*, *FD*, *DE* be the zone-circles e f g, h k l, p q r; *O* the pole *m n o*; *P* the pole *u v w*. Then (16)

$$\frac{\sin EFO}{\sin EFP}\,\frac{\sin DFP}{\sin DFO}=\frac{em+fn+go}{eu+fv+gw}\,\frac{hu+kv+lw}{hm+kn+lo},$$

$$\frac{\sin FEO}{\sin FEP}\,\frac{\sin DEP}{\sin DEO}=\frac{em+fn+go}{eu+fv+gw}\,\frac{pu+qv+rw}{pm+qn+ro}.$$

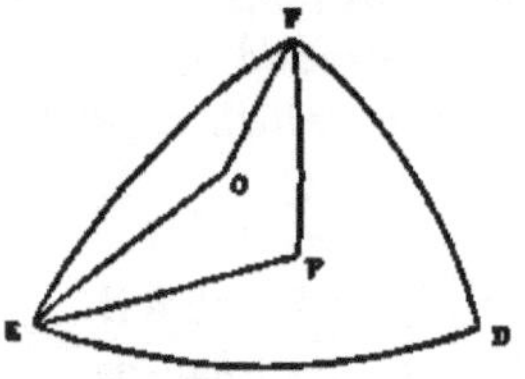

Let *m'n'o'* be the symbol of *O*, *u'v'w'* the symbol of *P*, when referred to the axes of the zone-circles *EF*, *FD*, *DE* as axes of the system of planes. Then (6) the new symbols of *EF*, *FD*, *DE* will be 1 0 0, 0 1 0, 0 0 1. Therefore (16)

$$\frac{\sin EFO}{\sin EFP}\,\frac{\sin DFP}{\sin DFO}=\frac{m'}{u'}\,\frac{v'}{n'},\qquad \frac{\sin FEO}{\sin FEP}\,\frac{\sin DEP}{\sin DEO}=\frac{m'}{u'}\,\frac{w'}{o'}.$$

Hence, equating the right-hand sides of equations having identical left-hand terms, we obtain two equations which are satisfied by making

$$m'=em+fn+go,\qquad u'=eu+fv+gw,$$

$$n'=hm+kn+lo,\qquad v'=hu+kv+lw,$$

$$o'=pm+qn+ro,\qquad w'=pu+qv+rw.$$

The coefficients of *u*, *v*, *w* are integers, therefore *u'*, *v'*, *w'*, the indices of *P* when referred to the axes of the zone-circles e f g, h k l, p q r as axes of the system of planes, will also be integers. Hence, the planes of the system are subject to the same law when referred to any three zone-axes, as when referred to their original axes.

22. Let D, E, F be the poles efg, hkl, pqr. Let EF, FD, DE meet the zone-circle mno in M, N, O, and the zone-circle uvw in U, V, W. Then (12)

$$\frac{\sin OD \sin WE}{\sin OE \sin WD} = \frac{\mathrm{m}e + \mathrm{n}f + \mathrm{o}g}{\mathrm{m}h + \mathrm{n}k + \mathrm{o}l} \, \frac{\mathrm{u}h + \mathrm{v}k + \mathrm{w}l}{\mathrm{u}e + \mathrm{v}f + \mathrm{w}g},$$

$$\frac{\sin ND}{\sin NF} \frac{\sin VF}{\sin VD} = \frac{\mathrm{m}e + \mathrm{n}f + \mathrm{o}g}{\mathrm{m}p + \mathrm{n}q + \mathrm{o}r} \, \frac{\mathrm{u}p + \mathrm{v}q + \mathrm{w}r}{\mathrm{u}e + \mathrm{v}f + \mathrm{w}g}.$$

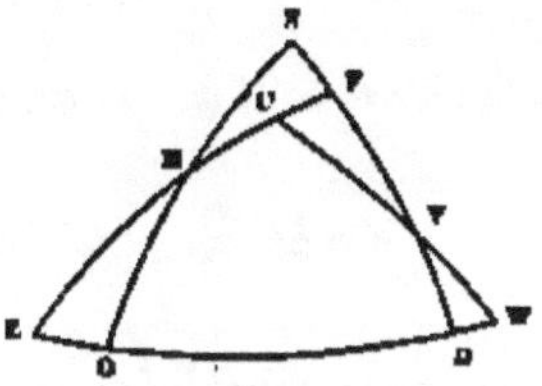

Let m′ n′ o′ be the symbol of the zone-circle MO, u′ v′ w′ the symbol of the zone-circle UW, when referred to the axes of the zone-circles EF, FD, DE as axes of the system of planes. Then (6), (7) the new symbols of D, E, F will be 1 0 0, 0 1 0, 0 0 1. Therefore (12)

$$\frac{\sin OD}{\sin OE} \frac{\sin WE}{\sin WD} = \frac{\mathrm{m}'}{\mathrm{n}'} \frac{\mathrm{v}'}{\mathrm{u}'}, \qquad \frac{\sin ND}{\sin NF} \frac{\sin VF}{\sin VD} = \frac{\mathrm{m}'}{\mathrm{o}'} \frac{\mathrm{w}'}{\mathrm{u}'}.$$

Hence, equating the right-hand sides of the equations having identical left-hand terms, we obtain two equations which are satisfied by making

$$\mathrm{m}' = e\mathrm{m} + f\mathrm{n} + g\mathrm{o}, \qquad \mathrm{u}' = e\mathrm{u} + f\mathrm{v} + g\mathrm{w},$$

$$\mathrm{n}' = h\mathrm{m} + k\mathrm{n} + l\mathrm{o}, \qquad \mathrm{v}' = h\mathrm{u} + k\mathrm{v} + l\mathrm{w},$$

$$\mathrm{o}' = p\mathrm{m} + q\mathrm{n} + r\mathrm{o}, \qquad \mathrm{w}' = p\mathrm{u} + q\mathrm{v} + r\mathrm{w}.$$

23. Let hkl, uvw be the symbols of the poles O, P, the parameters of the system of planes being a, b, c; $h'k'l'$, $u'v'w'$ the symbols of O, P when referred to the same axes, but with the parameters a', b', c'. Then (2)

$$\frac{a}{h}\cos XO = \frac{b}{k}\cos YO = \frac{c}{l}\cos ZO,$$

$$\frac{a'}{h'}\cos XO = \frac{b'}{k'}\cos YO = \frac{c'}{l'}\cos ZO,$$

$$\frac{a}{u}\cos XP = \frac{b}{v}\cos YP = \frac{c}{w}\cos ZP,$$

$$\frac{a'}{u'}\cos XP = \frac{b'}{v'}\cos YP = \frac{c'}{w'}\cos ZP.$$

Hence $hu' : h'u = kv' : k'v = lw' : l'w$. These equations are satisfied by making

$$u' = h'klu, \qquad v' = hk'lv, \qquad w' = hkl'w.$$

24. Let $h\,k\,l$ be the symbol of a pole, u v w that of a zone-circle, the parameters being a, b, c; $h'\,k'\,l'$, u′ v′ w′ the symbols of the same pole and zone-circle when referred to the same axes, but with the parameters a', b', c'.

Let $m\,n\,o$, $p\,q\,r$ be the symbols of any two poles in the zone-circle, the parameters being a, b, c; $m'\,n'\,o'$, $p'\,q'\,r'$ their symbols, the parameters being a', b', c'. Then (5)

$$\mathrm{u} = nr - oq, \quad \mathrm{v} = op - mr, \quad \mathrm{w} = mq - np,$$

$$\mathrm{u}' = n'r' - o'q', \quad \mathrm{v}' = o'p' - m'r', \quad \mathrm{w}' = m'q' - n'p'.$$

But (23) $m' = h'klm$, $n' = hk'ln$, $o' = hkl'o$, $p' = h'klp$, $q' = hk'lq$, $r' = hkl'r$. Substituting these values of m', n', o', p', q', r' in the expressions for u′, v′, w′, and rejecting the common factor $h\,k\,l$, we obtain

$$\mathrm{u}' = hk'l'\mathrm{u}; \qquad \mathrm{v}' = h'kl'\mathrm{v}, \qquad \mathrm{w}' = h'k'l\mathrm{w}.$$

25. Let K be the pole of the zone-circle u v w; P, Q, R poles of the great-circles KX, KY, KZ. The great-circles YP, ZP, ZQ, XQ, XR, YR make with the great-circles KX, KY, KZ six right-angled triangles having KX, KY, KZ for the sides opposite to their right angles. Hence,

$$\cos YP = \sin XKY \sin KY, \quad -\cos ZP = \sin ZKX \sin KZ,$$
$$\cos ZQ = \sin YKZ \sin KZ, \quad -\cos XQ = \sin XKY \sin KX,$$
$$\cos YR = \sin YKZ \sin KY, \quad -\cos XR = \sin ZKX \sin KX.$$

Since P, Q, R are poles of KX, KY, KZ, $\cos XP = 0$, $\cos YQ = 0$, $\cos ZR = 0$. KP, KQ, KR are quadrants, therefore P, Q, R are points in the zone-circle u v w. Hence (5)

$$vb \cos YP + wc \cos ZP = 0,$$
$$ua \cos XQ + wc \cos ZQ = 0,$$
$$ua \cos XR + vb \cos YR = 0.$$

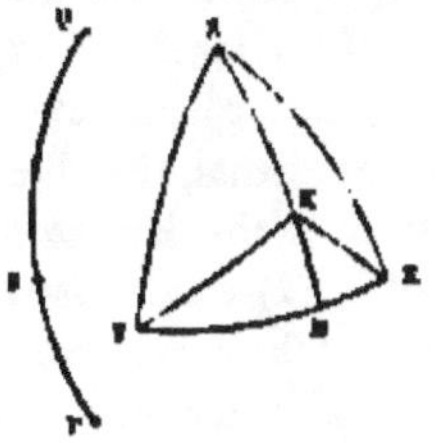

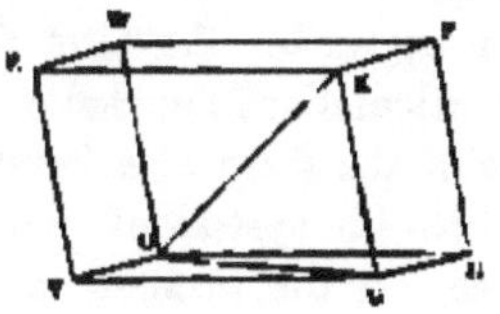

Therefore $ua \dfrac{\sin KX}{\sin YKZ} = vb \dfrac{\sin KY}{\sin ZKX} = wc \dfrac{\sin KZ}{\sin XKY}$.

Construct a parallelopiped UVW having OK, the axis of the zone, for a diagonal, and three of its edges OU, OV, OW coincident with the axes of the system of planes. Let KE, KF, KG be the edges respectively parallel to OU, OV, OW. The angles GOU, GOV are the segments into which UOV is divided by OG, the intersection of the planes WOK, UOV, and are therefore measured by the arcs NX, NY, N being the intersection of XY and KZ. Hence

$$\frac{OV}{OU} = \frac{\sin GOU}{\sin GOV} = \frac{\sin NX}{\sin NY} = \frac{\sin KX \sin ZKX}{\sin KY \sin YKZ} = \frac{vb}{ua}.$$

In like manner $\dfrac{OW}{OU} = \dfrac{cw}{ua}$.

Therefore $\dfrac{OU}{ua} = \dfrac{OV}{vb} = \dfrac{OW}{wc}$.

Or, the axis of the zone u v w is the diagonal of a parallelopiped, the edges of which coincide with the axes of the system of planes, and are equal to *ua*, *vb*, *wc* respectively.

26. Many natural substances, and many of the results of chemical operations, occur in the form of polyhedral solids. These, when broken, frequently separate in the directions of planes passing through any point within the solid, either parallel to certain planes of the solid, or making invariable angles with them. Solids of this description are called *crystals;* the planes by which they are bounded, their *faces;* and the planes in which they separate, their *cleavage planes.* It appears from accurate measurements of the mutual inclinations of the faces of a crystal, including under the term faces, its cleavage planes also, and from calculations founded on those measurements, that the positions of the faces of a crystal are subject to the law according to which the system of planes described in (1) was constructed. Hence, all the geometrical properties which have been established for such a system of planes, are also properties of the system of planes by which a crystal is bounded.

The angle between any two of the faces of a crystal will be measured by the plane angle between normals to the two faces, drawn towards the planes of the faces, from any point within the crystal, or by the arc of a great-circle of the sphere of projection joining the poles of the faces.

27. In many crystals axes may be discovered which make right angles with one another; in others, axes of which one makes right angles with each of the other two; and in others, axes making equal oblique angles with one another. In the crystals with equiangular axes, and in some of the crystals with rectangular axes, the parameters are all equal; and among the remaining crystals with rectangular axes, some which have two of the parameters equal. Upon these differences in the mutual inclinations of the axes, and in the relation betwen the parameters, is founded the arrangement of crystals in systems. The

different systems are further distinguished by the various kinds of symmetry observable in the distribution of the faces of the crystals belonging to them; for, if a face occur having the symbol $h\,k\,l$, it will generally be accompanied by the faces having for their symbols certain arrangements of $\pm h$, $\pm k$, $\pm l$ determined by laws peculiar to each system.

28. The figure consisting of a given face and the faces which, by the law of symmetry of the system of crystalization, are required to coexist with it, is called a *form*. The form consisting of the face $h\,k\,l$ and its coexistent faces, may be denoted by the symbol $\{h\,k\,l\}$. When, however, there is no danger of mistaking the form for a zone or a face having the same indices, the braces may be omitted.

Forms possessing all the faces required by the law of symmetry of the system to which they belong, are sometimes called *holohedral*, in order to distinguish them from peculiar forms of frequent occurrence, which are derived from holohedral forms by suppressing half of their faces according to certain laws, and are called *hemihedral*. The figure consisting of the faces of any number of forms is called a *combination* of those forms.

CHAPTER II.

CUBIC SYSTEM.

29. In the cubic system the axes make right angles with one another, and the parameters are all equal.

30. The form $h\,k\,l$ is contained by the faces having for their symbols the different arrangements of $\pm h$, $\pm k$, $\pm l$. These are:

$h\,k\,l$	$k\,l\,h$	$l\,h\,k$	$l\,k\,h$	$k\,h\,l$	$h\,l\,k$
$h\,\bar{k}\,\bar{l}$	$k\,\bar{l}\,\bar{h}$	$l\,\bar{h}\,\bar{k}$	$l\,\bar{k}\,\bar{h}$	$k\,\bar{h}\,\bar{l}$	$h\,\bar{l}\,\bar{k}$
$\bar{h}\,k\,\bar{l}$	$\bar{k}\,l\,\bar{h}$	$\bar{l}\,h\,\bar{k}$	$\bar{l}\,k\,\bar{h}$	$\bar{k}\,h\,\bar{l}$	$\bar{h}\,l\,\bar{k}$
$\bar{h}\,\bar{k}\,l$	$\bar{k}\,\bar{l}\,h$	$\bar{l}\,\bar{h}\,k$	$\bar{l}\,\bar{k}\,h$	$\bar{k}\,\bar{h}\,l$	$\bar{h}\,\bar{l}\,k$
$\bar{h}\,\bar{k}\,\bar{l}$	$\bar{k}\,\bar{l}\,\bar{h}$	$\bar{l}\,\bar{h}\,\bar{k}$	$\bar{l}\,\bar{k}\,\bar{h}$	$\bar{k}\,\bar{h}\,\bar{l}$	$\bar{h}\,\bar{l}\,\bar{k}$
$\bar{h}\,k\,l$	$\bar{k}\,l\,h$	$\bar{l}\,h\,k$	$\bar{l}\,k\,h$	$\bar{k}\,h\,l$	$\bar{h}\,l\,k$
$h\,\bar{k}\,l$	$k\,\bar{l}\,h$	$l\,\bar{h}\,k$	$l\,\bar{k}\,h$	$k\,\bar{h}\,l$	$h\,\bar{l}\,k$
$h\,k\,\bar{l}$	$k\,l\,\bar{h}$	$l\,h\,\bar{k}$	$l\,k\,\bar{h}$	$k\,h\,\bar{l}$	$h\,l\,\bar{k}$

When h, k, l are all different, the number of arrangements will be forty-eight; when any two indices are equal, it will be twenty-four; when two of the indices are equal, and the third is zero, it will be twelve; when all three indices are equal, it will be eight; and when two of the indices are zero, it will be six.

31. The form contained either by the faces of the form $h\,k\,l$ which have an odd number of positive indices, or by the faces which have an odd number of negative indices, is said to be hemihedral with inclined faces. It will be denoted by the symbol $\kappa\,h\,k\,l$, where $h\,k\,l$ is the symbol of any one of its faces. The symbols in the upper and lower halves of the table in (30) are those of the two half forms respectively.

32. The form contained either by the faces of the form $h\,k\,l$ having their indices in the order $h\,k\,l\,h\,k$, or by the faces having their indices in the order $l\,k\,h\,l\,k$, is said to be hemihedral with parallel faces. It will be denoted by the symbol $\pi\,h\,k\,l$, where $h\,k\,l$ is the symbol of any one of its faces. The symbols in the left and right halves of the table in (30) are those of the two half forms respectively.

33. Let A, B, C be the poles $1\,0\,0$, $0\,1\,0$, $0\,0\,1$ respectively; P the pole $h\,k\,l$. The axes make right angles with one another, therefore the sides of the triangle XYZ are quadrants, its angles are right angles, and X, Y, Z are poles of the arcs YZ, ZX, XY. But A, B, C are poles of YZ, ZX, XY, and they have no negative indices, therefore (3) A, B, C coincide with X, Y, Z respectively. Hence, the sides of the triangle ABC are quadrants, and its angles are right angles. The quadrantal triangles PAB, PBC give

$$(\cos BP)^2 = (\sin AP)^2 (\cos BAP)^2,$$

$$(\cos CP)^2 = (\sin AP)^2 (\cos CAP)^2.$$

Add, observing that $(\cos BAP)^2 + (\cos CAP)^2 = 1$, and that $(\cos AP)^2 + (\sin AP)^2 = 1$, and we obtain

$$(\cos AP)^2 + (\cos BP)^2 + (\cos CP)^2 = 1.$$

The parameters are all equal, and A, B, C coincide with X, Y, Z, therefore (2),

$$\frac{1}{h}\cos AP = \frac{1}{k}\cos BP = \frac{1}{l}\cos CP.$$

Hence
$$(\cos AP)^2 = \frac{h^2}{h^2+k^2+l^2},$$
$$(\cos BP)^2 = \frac{k^2}{h^2+k^2+l^2},$$
$$(\cos CP)^2 = \frac{l^2}{h^2+k^2+l^2}.$$

34. Let P, Q be the poles $h\,k\,l$, $p\,q\,r$ respectively.

$$\cos PQ = \cos AP \cos AQ + \sin AP \sin AQ \cos PAQ,$$
$$\cos PAQ = \cos BAP \cos BAQ + \sin BAP \sin BAQ,$$
$$\sin AP \cos BAP = \cos BP, \quad \sin AQ \cos BAQ = \cos BQ,$$
$$\sin AP \sin BAP = \cos CP, \quad \sin AQ \sin BAQ = \cos CQ.$$

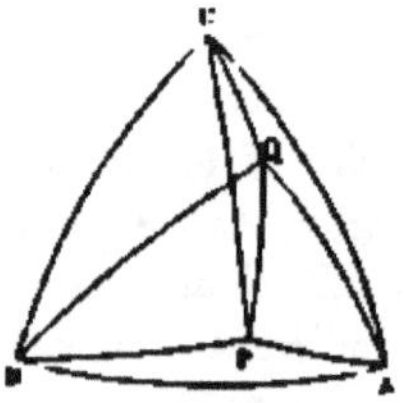

Hence
$$\cos PQ = \cos AP \cos AQ + \cos BP \cos BQ + \cos CP \cos CQ.$$

$$(\cos AP)^2 = \frac{h^2}{h^2+k^2+l^2}, \quad (\cos AQ)^2 = \frac{p^2}{p^2+q^2+r^2},$$
$$(\cos BP)^2 = \frac{k^2}{h^2+k^2+l^2}, \quad (\cos BQ)^2 = \frac{q^2}{p^2+q^2+r^2},$$
$$(\cos CP)^2 = \frac{l^2}{h^2+k^2+l^2}, \quad (\cos CQ)^2 = \frac{r^2}{p^2+q^2+r^2}.$$

Therefore
$$\cos PQ = \frac{hp+kq+lr}{\sqrt{(h^2+k^2+l^2)}\,\sqrt{(p^2+q^2+r^2)}}.$$

35. The quadrantal triangles BPC, CPA, APB give

$$\cos AP = \sin BP \cos ABP = \sin CP \cos ACP,$$

$$\cos BP = \sin CP \cos BCP = \sin AP \cos BAP,$$

$$\cos CP = \sin AP \cos CAP = \sin BP \cos CBP.$$

But $\frac{1}{h}\cos AP = \frac{1}{k}\cos BP = \frac{1}{l}\cos CP$. Hence

$$\tan BAP = \frac{l}{k}, \quad \tan CBP = \frac{h}{l}, \quad \tan ACP = \frac{k}{h}.$$

36. It appears from the expressions in (33), that if the symbols of two poles of the form $h\,k\,l$ differ only in the signs of h, they will be equidistant from the pole 0 1 0, and also equidistant from the pole 0 0 1. Therefore the arc joining the two poles will be bisected at right angles by the zone-circle passing through the poles 0 1 0, 0 0 1. Hence, the poles of the form $h\,k\,l$ are symmetrically situated with respect to the zone-circle passing through the poles 0 1 0, 0 0 1, and the two diametrically opposite poles. In like manner the poles of the form $h\,k\,l$ are symmetrically situated with respect to any one of the three zone-circles containing four poles of the form 1 0 0. It appears from (34) that, if the symbols of any two poles of the form $h\,k\,l$ differ only in the arrangement of the second and third indices, the poles will be equidistant from 1 1 1, and also from 1 1 1. Therefore the arc joining the two poles will be bisected at right angles by the zone-circle passing through the poles 1 1 1, 1 1 1, and the two opposite poles. In like manner the poles of the form $h\,k\,l$ are symmetrically situated with respect to any one of the six zone-circles containing four poles of the form 1 1 1.

The poles of a hemihedral form with inclined faces are symmetrically situated with respect to each of the six zone-circles containing the poles of the form 1 1 1.

The poles of a hemihedral form with parallel faces are symmetrically situated with respect to each of the three zone-circles containing the poles of the form 1 0 0.

37. If h be supposed the greatest, and l the least of three unequal indices h, k, l, the first of the annexed figures will represent the distribution of the poles of the form $h\,k\,l$ on one-

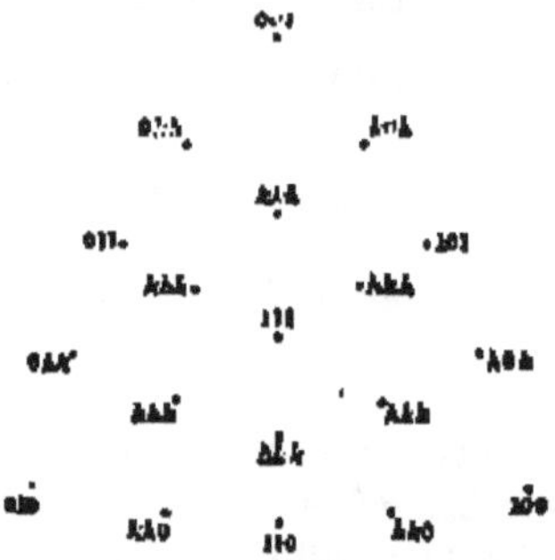

eighth of the sphere of projection. The second figure exhibits the poles of the forms obtained by making one of the indices zero, or by making two of them equal. Both figures show the poles of the forms 1 0 0, 1 1 1, and 1 1 0.

If the surface of the sphere be divided into eight triangles by the three zone-circles passing through the poles of the form 1 0 0, the poles of a hemihedral form with inclined faces will be found in four alternate triangles.

If the surface of the sphere be divided into twenty-four triangles by the six zone-circles passing through the poles of the form 1 1 1, the poles of a hemihedral form with parallel faces will be found in twelve alternate triangles.

38. The two hemihedral forms either with inclined or with parallel faces, derived from the same holohedral form, differ only in position. For, by turning the sphere of projection through a right angle, round a diameter joining any two opposite poles of the form 1 0 0, the poles of one of the two hemihedral forms derived from the same holohedral form, will change places with those of the other. But a combination of any two hemihedral forms derived from the forms $h\,k\,l$, $p\,q\,r$, when their poles fall in the same triangles formed by the system of zone-

circles passing through the poles of the form 1 0 0, or of the form 1 1 1, is essentially different from a combination of the hemihedral forms when their poles fall in different triangles.

39. The form 1 0 0 has six faces. Let F be the arc joining any two adjacent poles. Then $\cos F = 0$, therefore $F = 90°$. Hence the faces of the form 1 0 0 are respectively parallel to those of a cube.

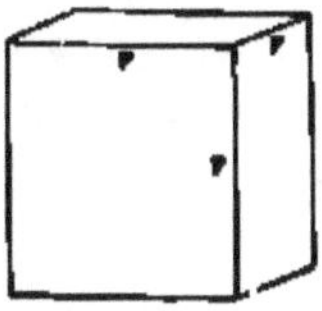

40. The form 1 1 1 has eight faces. Denoting by D the arc joining any two adjacent poles, we have $\cos D = \frac{1}{3}$. Therefore $D = 70° 31'·7$. Hence the faces of the form 1 1 1 are parallel to those of a regular octahedron.

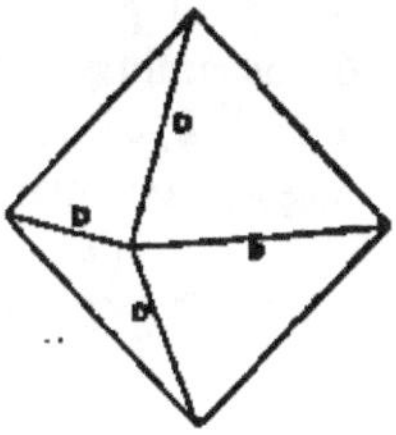

The cosine of the arc joining any pole of the form 1 1 1, and each of the adjacent poles of the form 1 0 0 is $\frac{1}{3}\sqrt{3}$. The corresponding arc is $54° 44'·1$.

41. Each of the forms κ 1 1 1, κ $\bar{1}\,\bar{1}\,\bar{1}$ has four faces. Let T be the arc joining any two adjacent poles. Then $\cos T = -\frac{1}{3}$, therefore $T = 109° 28'·3$. Hence each of the hemihedral forms is a regular tetrahedron.

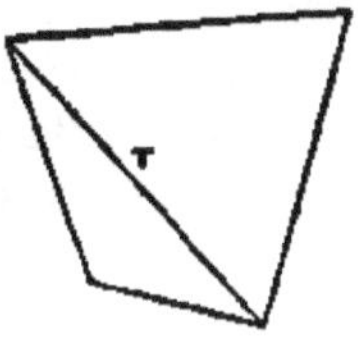

42. The form 0 1 1 has twelve faces. The arc joining any two adjacent poles being denoted by G, we have $\cos G = \frac{1}{2}$. Therefore $G = 60°$. The arc joining the poles of any two alternate faces, meeting at their acute angles, being denoted by D, $\cos D = 0$. Therefore $D = 90°$.

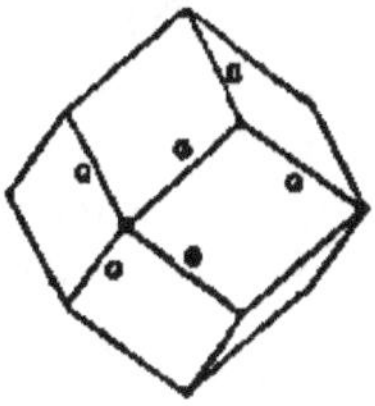

The arcs joining any pole of the form

0 1 1, and the two adjacent poles, the two opposite poles and the two remaining poles of the form 1 0 0, have for their cosines $\frac{1}{2}\sqrt{2}$, $-\frac{1}{2}\sqrt{2}$, 0, respectively. The corresponding angles are 45°, 135°, 90°. The arcs joining any pole of the form 0 1 1, and the two adjacent poles, the two opposite poles and the four remaining poles of the form 1 1 1, have for their cosines $\frac{1}{3}\sqrt{6}$, $-\frac{1}{3}\sqrt{6}$, 0. The corresponding angles are 35° 15′·85, 144° 44′·15, 90°.

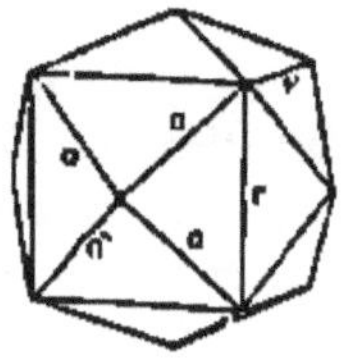

43. The form $h\,k\,0$ has twelve faces. Let the arc joining any two adjacent poles be F or G, according as they differ only in the order of h, k, or in the order of k, 0. Then, h being greater than k,

$$\cos F = \frac{2hk}{h^2+k^2}, \quad \cos G = \frac{h^2}{h^2+k^2}.$$

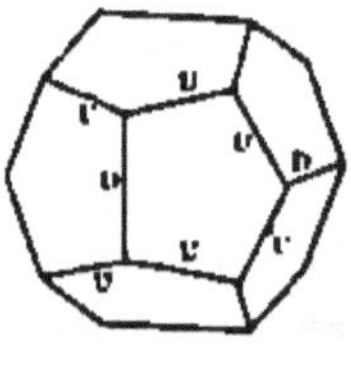

44. Each of the forms $\pi\,h\,k\,0$, $\pi\,0\,k\,h$ is contained by the alternate faces of the form $h\,k\,0$. Denoting by D the arc joining any two adjacent poles differing only in the signs of k, and by U the arc joining any two adjacent poles in the symbols of which the indices occupy different places, we have

$$\cos D = \frac{h^2-k^2}{h^2+k^2}, \quad \cos U = \frac{hk}{h^2+k^2}.$$

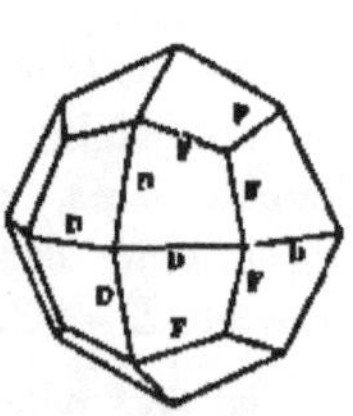

45. The form $h\,k\,k$ has twenty-four faces. Denoting the arc joining any two adjacent poles by D or F, according as the order of their indices is the same or different, h being greater than k, we have

$$\cos D = \frac{h^2}{h^2+2k^2}, \quad \cos F = \frac{2hk+k^2}{h^2+2k^2}.$$

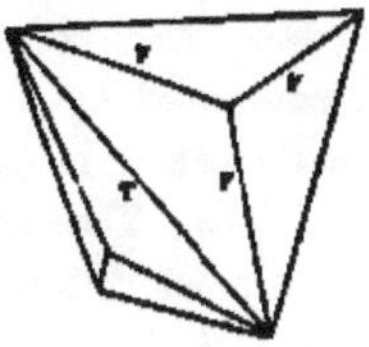

46. Each of the forms $\kappa\, h\, k\, k$, $\kappa\, \bar{h}\, \bar{k}\, \bar{k}$ is contained by the alternate triads of faces which meet in the edges F of the form $h\, k\, k$. Let T be the arc joining any two adjacent poles differing only in the signs of k. Then

$$\cos T = \frac{h^2 - 2k^2}{h^2 + 2k^2}.$$

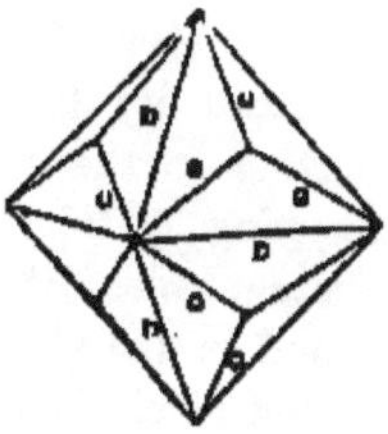

47. The form $h\, h\, k$ has twenty-four faces. Denoting the arc joining any two adjacent poles by D or G, according as the order of the indices is the same or different, h being greater than k, we have

$$\cos D = \frac{2h^2 - k^2}{2h^2 + k^2}, \quad \cos G = \frac{h^2 + 2hk}{2h^2 + k^2}.$$

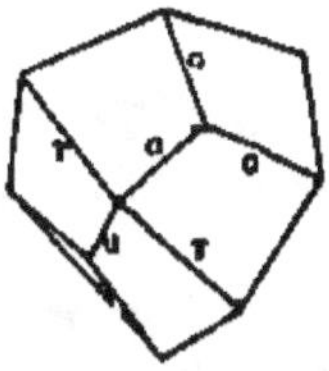

48. Each of the forms $\kappa\, h\, h\, k$, $\kappa\, \bar{h}\, \bar{h}\, \bar{k}$ is contained by the alternate triads of faces which meet in the edges G of the form $h\, h\, k$. Denoting by T the arc joining any two adjacent poles differing in the order of the indices, and in the signs of two of them, we have

$$\cos T = \frac{h^2 - 2hk}{2h^2 + k^2}.$$

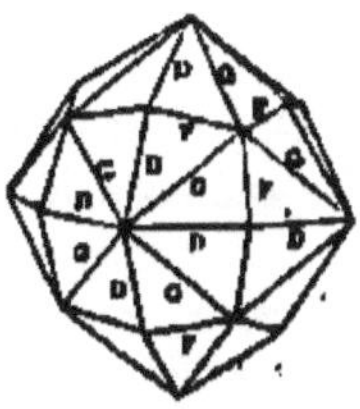

49. The form $h\, k\, l$ has forty-eight faces. Denoting by D, F, G the arcs joining adjacent poles differing only in the signs of l, in the order of h, k, and in the order of k, l respectively, h being greater, and l less than k, we have

$$\cos D = \frac{h^2 + k^2 - l^2}{h^2 + k^2 + l^2}, \quad \cos F = \frac{2hk + l^2}{h^2 + k^2 + l^2}, \quad \cos G = \frac{h^2 + 2kl}{h^2 + k^2 + l^2}.$$

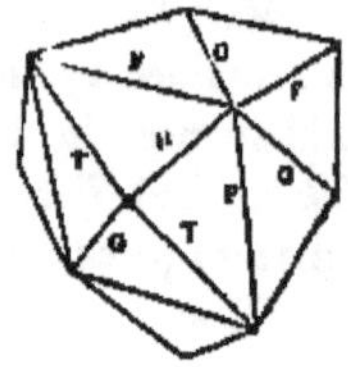

50. Each of the forms $\kappa\, h\, k\, l$, $\kappa\, \bar{h}\, \bar{k}\, \bar{l}$ is contained by the alternate groups of six faces meeting in the edges F, G of the form $h\, k\, l$. Let T be the arc joining any two adjacent poles differing only in the order and signs of k, l. Then

$$\cos T = \frac{h^2 - 2kl}{h^2 + k^2 + l^2}.$$

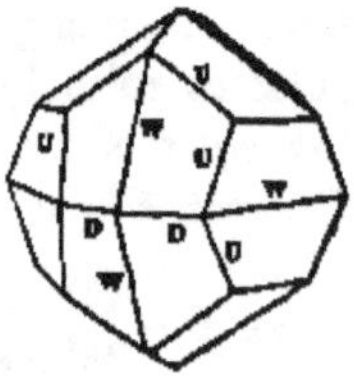

51. Each of the forms $\pi\, h\, k\, l$, $\pi\, l\, k\, h$ is contained by the alternate pairs of faces meeting in the edges D of the form $h\, k\, l$. Denoting by W, U the arcs joining any two adjacent poles differing only in the signs of k, and in the places occupied by the several indices, respectively, we have

$$\cos W = \frac{h^2 - k^2 + l^2}{h^2 + k^2 + l^2}, \quad \cos U = \frac{kl + lh + hk}{h^2 + k^2 + l^2}.$$

52. The cleavages are usually parallel to the faces of one or more of the forms 1 0 0, 1 1 1, 0 1 1.

53. If we have given the arc joining any two poles, not opposite to one another, of one of the forms $h\, k\, 0$, $h\, k\, k$, $h\, h\, k$, the expression for its cosine, in terms of the indices of the poles, will supply an equation from which the ratio of the indices may be deduced.

54. If we have given the arcs joining any pole of the form $h\, k\, l$, and each of two other poles of the same form, no two of the poles being opposite to one another, the expressions for their cosines, in terms of the indices of the poles, will supply two equations from which the ratios of the indices may be found.

CHAPTER III.

PYRAMIDAL SYSTEM.

55. In the pyramidal system the axes make right angles with one another, and the parameters a, b are equal.

56. The form $h\,k\,l$ consists of the faces which have for their symbols the different arrangements of $\pm h$, $\pm k$, $\pm l$, in which l holds the last place. These are:

$$\begin{array}{cccc} h\,k\,l & \bar{h}\,k\,\bar{l} & \bar{k}\,h\,l & \bar{k}\,\bar{h}\,\bar{l} \\ \bar{h}\,\bar{k}\,l & h\,\bar{k}\,\bar{l} & k\,\bar{h}\,l & k\,h\,\bar{l} \\ \bar{k}\,h\,\bar{l} & k\,h\,l & \bar{h}\,\bar{k}\,\bar{l} & \bar{h}\,k\,l \\ k\,\bar{h}\,\bar{l} & \bar{k}\,\bar{h}\,l & h\,k\,\bar{l} & h\,\bar{k}\,l \end{array}$$

When h and k are different, and l is finite, the number of faces will be sixteen; when one of the indices is zero, or when $h = k$, the number will be eight; when l is zero, and $h = k$, or one of the indices h, k is zero, the number of faces will be four; and when h and k are zero it will be two.

57. The form contained either by the faces of the form $h\,k\,l$ which have an odd number of positive indices, or by the faces which have an odd number of negative indices, is said to be hemihedral with inclined faces, and will be denoted by the

symbol $\kappa\, h\, k\, l$ where $h\, k\, l$ is the symbol of any one of its faces. The left and right halves of the table contain the symbols of the two half forms respectively.

58. A second hemihedral form with inclined faces, contained by the faces of the form $h\, k\, l$ in which the order of h, k changes with the sign of l, will be denoted by the symbol $\lambda\, h\, k\, l$, where $h\, k\, l$ is the symbol of any one of its faces. The first and fourth columns of the table contain the symbols of the faces of one half form, the second and third columns those of the other half form.

59. The form consisting of the faces of the form $h\, k\, l$ in which the order of h, k is the same or different according as h, k have the same or different signs, is said to be hemihedral with parallel faces, and will be denoted by the symbol $\pi\, h\, k\, l$, where $h\, k\, l$ is the symbol of any one of its faces. The first and third columns of the table contain the symbols of one half form, the second and fourth those of the other half form.

60. The form contained by the faces of the form $h\, k\, l$, in which the order of the indices h, k is the same or different according as an odd number of the indices are positive or negative, is said to be hemihedral with asymmetric faces, and will be denoted by the symbol $\alpha\, h\, k\, l$, where $h\, k\, l$ is the symbol of any one of its faces. The upper and lower halves of the table contain the symbols of the two half forms respectively.

61. Let a, a, c be the parameters; A, B, C the poles 1 0 0, 0 1 0, 0 0 1 respectively; P the pole $h\, k\, l$. The axes make right angles with one another, therefore the sides of the triangle XYZ are quadrants, its angles are right angles, and X, Y, Z are the poles of YZ, ZX, XY. But A, B, C are poles of YZ, ZX, XY, and they have no negative indices, therefore (8) A, B, C coincide with X, Y, Z respectively. Hence, the sides of the triangle ABC are quadrants, and its angles are right angles. The quadrantal triangles PBC, PCA, PAB give

$$\cos AP = \sin BP \cos ABP = \sin CP \cos ACP,$$
$$\cos BP = \sin CP \cos BCP = \sin AP \cos BAP,$$
$$\cos CP = \sin AP \cos CAP = \sin BP \cos CBP.$$
$$\cot AP = \tan BCP \cos BAP = \tan CBP \cos CAP,$$
$$\cot BP = \tan CAP \cos CBP = \tan ACP \cos ABP,$$
$$\cot CP = \tan ABP \cos ACP = \tan BAP \cos BCP.$$

Also, since A, B, C coincide with X, Y, Z,

$$\frac{a}{h} \cos AP = \frac{a}{k} \cos BP = \frac{c}{l} \cos CP.$$

Hence, substituting in the preceding equations the values of $\cos AP$, $\cos BP$, $\cos CP$ given above, we obtain

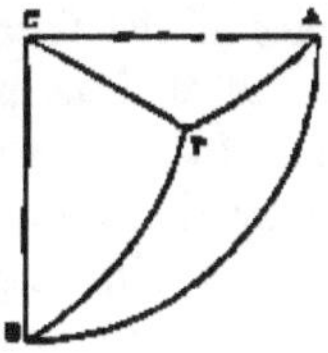

$$\tan BAP = \frac{l}{k} \frac{a}{c},$$
$$\tan ABP = \frac{l}{h} \frac{a}{c},$$
$$\tan ACP = \frac{k}{h}.$$

62. Let E be the arc joining the poles 0 0 1, 1 0 1. Then E measures the angle it subtends at B. Therefore the second of the preceding equations gives $\tan E = c : a$. Hence

$$\tan BAP = \frac{l}{k} \cot E, \quad \tan ABP = \frac{l}{h} \cot E, \quad \tan ACP = \frac{k}{h}.$$

$$\cot AP = \frac{h}{k} \cos BAP = \frac{h}{l} \tan E \cos CAP,$$
$$\cot BP = \frac{k}{l} \tan E \cos CBP = \frac{k}{h} \cos ABP,$$
$$\cot CP = \frac{l}{h} \cot E \cos ACP = \frac{l}{k} \cot E \cos BCP$$
$$(\tan CP)^2 = \frac{h^2 + k^2}{l^2} (\tan E)^2.$$

63. Since $\tan E = c : a$, E may be taken for the element of a crystal belonging to the pyramidal system.

64. The poles of the form 1 1 0 bisect the arcs joining any two adjacent poles of the form 1 0 0. For the poles of the forms 1 0 0, 1 1 0 are all in one zone-circle; the arc joining the poles 1 0 0, 0 1 0 is a quadrant; and (62) the arc joining the pole 1 0 0, and any pole of the form 1 1 0, having for its cotangent either 1 or −1, is an odd multiple of 45°.

65. It appears from the expressions in (62) that the arcs joining the poles of the form $h\,k\,l$, and the nearest of the two poles of the form 0 0 1, are all equal; and that the angles subtended at either pole of the form 0 0 1 by the arcs joining any pole of the form $h\,k\,l$, and the nearest pole of the form 1 0 0, are all equal. Hence, the poles of the form $k\,h\,l$ are symmetrically situated with respect to each of the five zone-circles containing poles of any two of the three forms 0 0 1, 1 0 0, 1 1 0.

The poles of the form $\kappa\,h\,k\,l$ are symmetrically situated with respect to each of the two zone-circles drawn through the poles of the form 0 0 1, and those of the form 1 1 0.

The poles of the form $\lambda\,h\,k\,l$ are symmetrically situated with respect to each of the two zone-circles through the poles of the form 0 0 1, and those of the form 1 0 0.

The poles of the form $\pi\,h\,k\,l$ are symmetrically situated with respect to the zone-circle containing the poles of the form 1 0 0.

66. If h be supposed greater than k, the annexed figure will represent the arrangement of the poles of the forms $h\,k\,l$, $h\,h\,l$, $h\,k\,0$, $h\,0\,l$, 1 1 0, 1 0 0, 0 0 1 on the surface of the sphere of projection.

If the surface of the sphere be divided into eight triangles by zone-circles passing through the poles of the forms 0 0 1, 1 0 0, the poles of the form $\kappa\,h\,k\,l$ will be found in four alternate triangles.

If the surface of the sphere be divided into eight triangles by zone-circles passing through the poles of the forms 0 0 1, 1 1 0, the poles of the form $\lambda\, h\, k\, l$ will be found in four alternate triangles.

If the surface of the sphere be divided into eight lunes by zone-circles passing through the poles of the form 0 0 1, and those of the forms 1 0 0, 1 1 0, the poles of the form $\pi\, h\, k\, l$ will be found in four alternate lunes.

The poles of the form $a\, h\, k\, l$ are eight alternate poles of the form $h\, k\, l$.

67. Any two hemihedral forms with inclined or with parallel faces, derived from the same holohedral form, differ only in position. For, by making the sphere of projection revolve through a right angle round a diameter joining the poles of the form 0 0 1, the poles of $\kappa\, h\, k\, l$ and $\lambda\, h\, k\, l$ will change places with those of $\kappa\, h\, k\, \bar{l}$ and $\lambda\, k\, h\, l$ respectively; and by making the sphere revolve through two right angles round a diameter joining any two opposite poles of the form 1 0 0, or of the form 1 1 0, the poles of $\pi\, h\, k\, l$ will change places with those of $\pi\, k\, h\, l$. The two forms $a\, h\, k\, l$, $a\, k\, h\, l$ are essentially different.

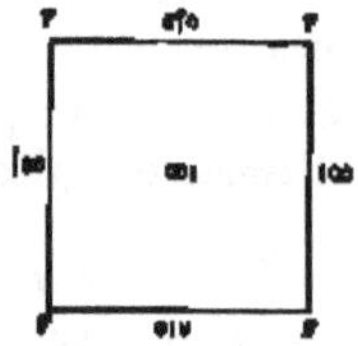

68. The form 0 0 1 has the two parallel faces 0 0 1, 0 0 $\bar{1}$.

69. The form 1 0 0 has four faces. Let F be the arc joining any two adjacent poles. Then $F = 1\,0\,0, 0\,1\,0$, and $\cot F = 0$. Therefore $F = 90^\circ$.

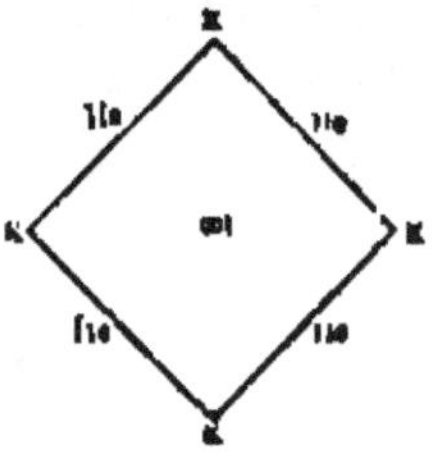

70. The form 1 1 0 has four faces. Let K be the arc joining any two adjacent poles. Then $\frac{1}{2}K = 1\,0\,0, 1\,1\,0$, and $\cot \frac{1}{2}K = 1$. Therefore $K = 90^\circ$.

In a combination of the forms 1 0 0 and 1 1 0, all the faces are in one zone, and any face of one form makes angles of 45° with the adjacent faces of the other form. The arc joining a pole of the form 0 0 1, and any pole of either of the forms 1 0 0, 1 1 0, is a quadrant. Therefore, in combinations of the form 0 0 1 with the forms 1 0 0, 1 1 0, the faces of the form 0 0 1 make right angles with those of the forms 1 0 0, 1 1 0.

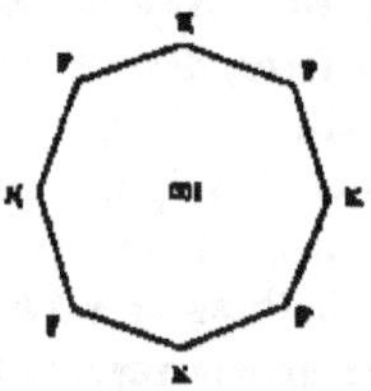

71. The form $h\,k\,0$ has eight faces in one zone. Let K be the arc joining any two adjacent poles differing in the signs of k; F the arc joining any two adjacent poles differing in the order of the indices h, k. Then $\frac{1}{2}K = 1\,0\,0, h\,k\,0$. Whence

$$\tan \tfrac{1}{2}K = \frac{k}{h}, \quad F = 90^\circ - K.$$

The arc joining a pole of the form 0 0 1, and any pole of the form $h\,k\,0$, is a quadrant. Therefore, in a combination of the forms 0 0 1, $h\,k\,0$, the faces of one form make right angles with those of the other form.

72. Each of the forms $\pi\, h\, k\, 0$, $\pi\, k\, h\, 0$, consists of the alternate faces of the form $h\, k\, 0$. Any two adjacent faces make right angles with one another.

73. The form $h\, 0\, l$ has eight faces. Let L be the arc joining any two adjacent poles differing in the signs of l; F the arc joining any two adjacent poles in the symbols of which l has the same sign. Then $90° - \frac{1}{2}L = 0\,0\,1, h\,0\,l$, and F subtends an angle of $90°$ at the pole $0\,0\,1$. Hence

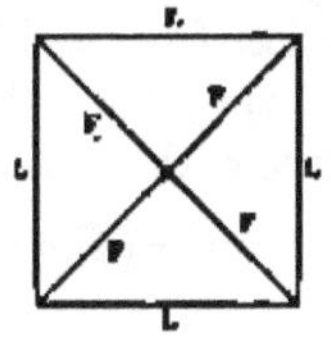

$$\tan \tfrac{1}{2}L = \frac{l}{h} \cot E, \quad \cos F = (\sin \tfrac{1}{2}L)^2.$$

74. Each of the forms $\lambda\, h\, 0\, l$, $\lambda\, 0\, h\, l$, is contained by the alternate faces of the form $h\, 0\, l$. Let U be the arc joining any two poles in which l has the same sign; V the arc joining any two poles in which l has different signs. Then

$$U = 180° - L, \quad V = 180° - F.$$

75. The form $h\, h\, l$ has eight faces. Let K be the arc joining any two adjacent poles in which l has the same sign, L the arc joining any two adjacent poles in which l has different signs. Then $90° - \frac{1}{2}L = 0\,0\,1, h\,h\,l$, and K subtends an angle of $90°$ at the pole $0\,0\,1$. Hence

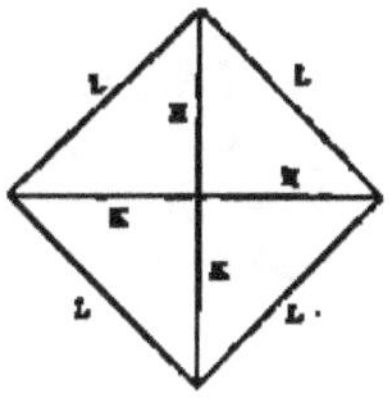

$$\tan \tfrac{1}{2}L = \frac{l}{h} \cot E \cos 45°, \quad \cos K = (\sin \tfrac{1}{2}L)^2.$$

76. Each of the forms $\kappa\, h\, h\, l$, $\kappa\, h\, h\, l$, consists of the alternate faces of the form $h\, h\, l$. Let W be the arc joining any two poles in which l has the same sign; T the arc joining any two poles in which l has different signs. Then

$$W = 180 - L, \quad T = 180° - K.$$

77. The form $h\,k\,l$ has sixteen faces. Let K, L be the arcs joining any two adjacent poles differing in the signs of k, l respectively; F the arc joining any two adjacent poles differing in the order of the indices h, k; and let ϕ be the angle which the arc joining the poles $1\,0\,0$, $h\,k\,l$, subtends at the pole $0\,0\,1$. Then

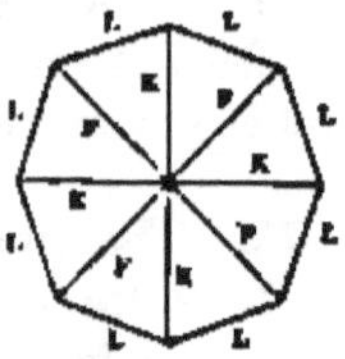

$$90^\circ - \tfrac{1}{2}L = 0\,0\,1, h\,k\,l, \quad 90^\circ - \tfrac{1}{2}K = 0\,1\,0, h\,k\,l. \quad \text{Hence}$$

$$\tan\phi = \frac{k}{h}, \quad \tan\tfrac{1}{2}L = \frac{l}{h}\cot E\cos\phi,$$

$$\sin\tfrac{1}{2}K = \cos\tfrac{1}{2}L\sin\phi, \quad \sin\tfrac{1}{2}F = \cos\tfrac{1}{2}L\sin(\tfrac{1}{4}\pi - \phi).$$

78. Each of the forms $\lambda\,h\,k\,l$, $\lambda\,k\,h\,l$, consists of the alternate pairs of faces of the form $h\,k\,l$ which meet in the edges K. Let H be the arc joining any two poles differing only in the signs of h; V the arc joining any two poles differing only in the order of h, k, and in the signs of l. Then

$$90^\circ - \tfrac{1}{2}H = 1\,0\,0, h\,k\,l, \quad \tfrac{1}{2}V = 1\,1\,0, h\,k\,l.$$

Hence

$$\sin\tfrac{1}{2}H = \cos\tfrac{1}{2}L\cos\phi, \quad \cos\tfrac{1}{2}V = \cos\tfrac{1}{2}L\cos(\tfrac{1}{4}\pi - \phi).$$

79. Each of the forms $\kappa\,h\,k\,l$, $\kappa\,h\,k\,\bar{l}$, consists of the alternate pairs of faces of the form $h\,k\,l$ which meet in the edges F. Let T be the arc joining any two poles differing only in the signs of k and l; G the arc joining any two poles differing only in the signs and order of h and k. Then

$$T = 180^\circ - H, \quad G = 180^\circ - V.$$

80. Each of the forms $\pi\,h\,k\,l$, $\pi\,k\,h\,l$, consists of the alternate pairs of faces of the form $h\,k\,l$ which meet in the edges L. Let M be the arc joining any two alternate poles of the form $h\,k\,l$, equidistant from the pole $0\,0\,1$. The angle subtended by M, at the pole $0\,0\,1$, will be 90°. Hence $\cos M = (\sin\tfrac{1}{2}L)^2$.

81. Each of the forms $a\,h\,k\,l$, $a\,k\,h\,l$, consists of the alternate faces of the form $h\,k\,l$. The arcs joining the adjacent poles

in the symbols of which l has the same sign, the signs of k are different, and the order of h, k different, are M, T, V respectively.

82. The principal cleavages are parallel to the faces of one or more of the forms 0 0 1, 1 0 0, 1 1 0, h 0 l, h h l.

83. Let C be the pole 0 0 1; P, Q any two adjacent poles of either of the forms $h\,h\,l$, $p\,0\,r$, equidistant from C; and let the arc PQ contain S a pole of the other form. Then CS will bisect the right angle PCQ, and the angle CSP will be a right angle. Whence, $\tan CS = \cos 45° \tan CP$.

84. Let A, B, C be the poles 1 0 0, 0 1 0, 0 0 1 respectively; P the pole $h\,k\,l$; Q the pole $p\,q\,r$. Then (62),

$$\cot AP = \frac{h}{k} \cos BAP = \frac{h}{l} \tan E \cos CAP,$$

$$\cot AQ = \frac{p}{q} \cos BAQ = \frac{p}{r} \tan E \cos CAQ.$$

Let Q be in the zone-circle AP. Then $BAQ = BAP$, and $CAQ = CAP$. Therefore

$$\frac{h}{p} \frac{\tan AP}{\tan AQ} = \frac{k}{q} = \frac{l}{r}.$$

In like manner, when Q is in the zone-circle BP,

$$\frac{k}{q} \frac{\tan BP}{\tan BQ} = \frac{l}{r} = \frac{h}{p}.$$

Also, when Q is in the zone-circle CP,

$$\frac{l}{r} \frac{\tan CP}{\tan CQ} = \frac{h}{p} = \frac{k}{q}.$$

85. Let C be the pole 0 0 1; P, Q the poles $h\,k\,l$, $p\,q\,r$ respectively. Then (62),

$$(\tan CP)^2 = \frac{h^2 + k^2}{l^2} (\tan E)^2,$$

$$(\tan CQ)^2 = \frac{p^2 + q^2}{r^2} (\tan E)^2.$$

Therefore $$\frac{l^2}{h^2+k^2}(\tan CP)^2 = \frac{r^2}{p^2+q^2}(\tan CQ)^2.$$

86. Let A, B, C be the poles 1 0 0, 0 1 0, 0 0 1 respectively; P, Q any two poles the symbols of which are given. Then, knowing E, and the symbols of P, Q, we can find CP, CQ, ACP, ACQ by (62). Hence, knowing CP, CQ and PCQ, the arc PQ can be found.

Or, having found the angles which CP, CQ subtend at one of the poles A, B, and the arcs joining this pole, and P, Q respectively, we have two sides and the included angle, from which the third side PQ may be found.

87. If the arc joining any two poles of the form $h\,k\,0$, not being either a quadrant or a semicircle, or the arc joining any two poles not opposite to one another, of either of the forms $h\,0\,l$, $h\,h\,l$, be given; the given arc, or its supplement, will be one of the arcs F, K, L (71), (73), (75). Hence an equation is obtained from which, knowing E, the ratio of the indices of the form may be found.

88. If we have given the arcs joining any pole of the form $h\,k\,l$, and each of two other poles of the same form, no two of the three poles being opposite to one another, the given arcs, or their supplements, will be two of the arcs H, K, L, F, V, M (77), (78), (80). Therefore two equations are obtained from which, knowing E, the ratios of the indices of the form may be found.

89. When the last index in the symbol of a form is finite, the arc joining any two poles not opposite to one another, or its supplement, is one of the arcs H, K, L, F, V, M. Therefore, if this arc and the symbol of the form be given, $\tan E$ may be found from the equations in (77), (78) or (80).

90. Let A, B, C be the poles 1 0 0, 0 1 0, 0 0 1 respectively; R, S any two poles in a zone-circle containing C; P the

intersection of RS and AB, PR being less than PS; p q r the symbol of any zone-circle, except RS, passing through R; $u\,v\,w$ the symbol of S. Then CP is a quadrant, and the symbol of AB, a zone-circle passing through P, is 0 0 1, therefore (20),

$$\sin(2PR + RS) = (2i - 1)\sin RS,$$

where
$$i = \frac{rw}{pu + qv + rw}.$$

Having found CR or CS by means of this equation, $\tan E$ is given by (62).

91. Let A, B, C be the poles 1 0 0, 0 1 0, 0 0 1 respectively; R, S any two poles not opposite to one another. Let RS meet AB in P, PR being less than PS, and let $s\,\bar{t}\,0$ be the symbol of P. Let M be the pole $t\,s\,0$; Q the intersection of RS and CM. Then $\tan ACM = -\cot ACP$, therefore PM is a quadrant. But CP is a quadrant, therefore PQ is a quadrant. The symbol of AB, a zone-circle passing through P, is 0 0 1. Let p q r be the symbol of any zone-circle passing through R, except RS. The numerical values of the indices of Q can be readily found from those of R and S, and the relation between the indices of P and M. Let $h\,k\,l$ be the symbol of Q, $u\,v\,w$ that of S. The arc PQ is a quadrant, therefore (20),

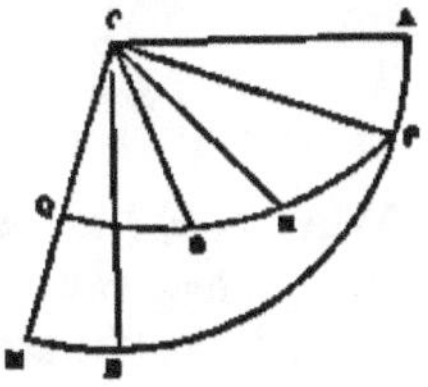

$$\sin(2PR + RS) = (2i - 1)\sin RS,$$

where
$$i = \frac{w}{l}\,\frac{ph + qk + rl}{pu + qv + rw}.$$

Having found PR or PS by means of this equation, and $\tan PCR$ or $\tan PCS$, we have

$$\cos PR = \cos PCR \sin CR, \quad \cos PS = \cos PCS \sin CS.$$

Hence, knowing CR or CS, $\tan E$ is given by (62).

CHAPTER IV.

RHOMBOHEDRAL SYSTEM.

92. In the rhombohedral system the axes make equal angles with one another, and the parameters are all equal.

93. The form $h\,k\,l$ consists of the faces which have for their symbols the different arrangements of $+h$, $+k$, $+l$, together with those of $-h$, $-k$, $-l$. These are

$h\,k\,l$	$l\,k\,h$	$\bar{h}\,\bar{k}\,\bar{l}$	$\bar{l}\,\bar{k}\,\bar{h}$
$k\,l\,h$	$k\,h\,l$	$\bar{k}\,\bar{l}\,\bar{h}$	$\bar{k}\,\bar{h}\,\bar{l}$
$l\,h\,k$	$h\,l\,k$	$\bar{l}\,\bar{h}\,\bar{k}$	$\bar{h}\,\bar{l}\,\bar{k}$

When h, k, l are all different, the number of faces will be twelve. When two of the indices are equal, or when they are 1, 0, -1, it will be six. When all three indices are equal, it will be two.

94. The form consisting either of the faces having for their symbols the different arrangements of $+h$, $+k$, $+l$, or of the faces having for their symbols the different arrangements of $-h$, $-k$, $-l$ is said to be hemihedral with inclined faces. It will be denoted by the symbol $\kappa\,h\,k\,l$, where $h\,k\,l$ is the symbol of any one of its faces. The left and right halves of the

table in (93) contain the symbols of the faces of the two half forms respectively.

95. The form consisting either of the faces of the form $h\,k\,l$ which have their indices in the order $h\,k\,l\,h\,k$, or of the faces which have their indices in the order $l\,k\,h\,l\,k$, is said to be hemihedral with parallel faces. It will be denoted by the symbol $\pi\,h\,k\,l$, where $h\,k\,l$ is the symbol of any one of its faces. The symbols of the faces of one half-form are contained in the first and third columns of the table in (93), those of the other in the second and fourth columns.

96. The form consisting either of the faces of the form $h\,k\,l$ having for their symbols the arrangements of $+h$, $+k$, $+l$ which stand in the order $h\,k\,l\,h\,k$, and those of $-h$, $-k$, $-l$ which stand in the order $l\,k\,h\,l\,k$, or of the faces having for their symbols the arrangements of $+h$, $+k$, $+l$ which stand in the order $l\,k\,h\,l\,k$, and those of $-h$, $-k$, $-l$ which stand in the order $h\,k\,l\,h\,k$, is said to be hemihedral with asymmetric faces, and will be denoted by the symbol $\alpha\,h\,k\,l$, where $h\,k\,l$ is the symbol of any one of its faces. The first and fourth columns of the table in (93) contain the symbols of the faces of one half-form; the second and third columns those of the other half-form.

97. Let O be the pole 1 1 1; P the pole $h\,k\,l$. Since the parameters are equal, and O is the pole 1 1 1, we shall have

$$\cos XO = \cos YO = \cos ZO, \text{ and } XO = YO = ZO.$$

The axes make equal angles with one another, therefore

$$YZ = ZX = XY.$$

Hence, YOZ, ZOX, XOY are each 120°. Therefore

$$\cos YOP = \cos 120^\circ \cos XOP + \sin 120^\circ \sin XOP,$$

$$\cos ZOP = \cos 120^\circ \cos XOP - \sin 120^\circ \sin XOP.$$

Hence, observing that $2\sin 120° = \sqrt{3}$, and $2\cos 120° = -1$,

$$\cos YOP - \cos ZOP = \sin XOP\sqrt{3},$$

$$\cos XOP + \cos YOP + \cos ZOP = 0.$$

$$\cos XP = \cos XO \cos OP + \sin XO \sin OP \cos XOP,$$

$$\cos YP = \cos YO \cos OP + \sin YO \sin OP \cos YOP,$$

$$\cos ZP = \cos ZO \cos OP + \sin ZO \sin OP \cos ZOP.$$

Hence $\sin XO \sin OP \sin XOP\sqrt{3} = \cos YP - \cos ZP,$

$$3\sin XO \sin OP \cos XOP = 2\cos XP - \cos YP - \cos ZP,$$

$$3\cos XO \cos OP = \cos XP + \cos YP + \cos ZP.$$

But $$\frac{1}{h}\cos XP = \frac{1}{k}\cos YP = \frac{1}{l}\cos ZP.$$

Hence $$\tan XOP = \frac{(k-l)\sqrt{3}}{2h-k-l},$$

$$\tan XO \tan OP \cos XOP = \frac{2h-k-l}{h+k+l}.$$

Similarly $$\tan YOP = \frac{(l-h)\sqrt{3}}{2k-l-h},$$

$$\tan YO \tan OP \cos YOP = \frac{2k-l-h}{h+k+l}.$$

And $$\tan ZOP = \frac{(h-k)\sqrt{3}}{2l-h-k},$$

$$\tan ZO \tan OP \cos ZOP = \frac{2l-h-k}{h+k+l}.$$

Also $$(\tan XO)^2 (\tan OP)^2 = 2\frac{(k-l)^2 + (l-h)^2 + (h-k)^2}{(h+k+l)^2}.$$

98. Let A, B, C be the poles 1 0 0, 0 1 0, 0 0 1 respectively. Then (97) $\tan XOA = 0$, $\tan YOB = 0$, $\tan ZOC = 0$, $\tan XO \tan OA = 2$, $\tan YO \tan OB = 2$, $\tan ZO \tan OC = 2$. Hence A, B, C are in the great circles OX, OY, OZ, and $OA = OB = OC$. Let $OA = D$. The expressions in (97) become

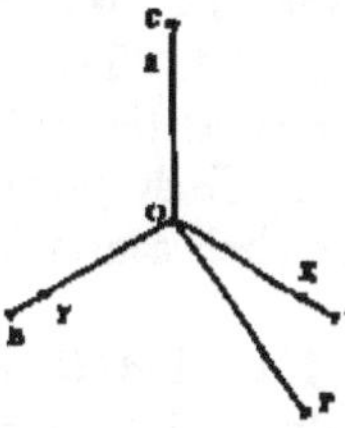

$$\tan AOP = \frac{(k-l)\sqrt{3}}{2h-k-l}, \quad 2\tan OP \cos AOP = \frac{2h-k-l}{h+k+l}\tan D,$$

$$\tan BOP = \frac{(l-h)\sqrt{3}}{2k-h-l}, \quad 2\tan OP \cos BOP = \frac{2k-h-l}{h+k+l}\tan D,$$

$$\tan COP = \frac{(h-k)\sqrt{3}}{2l-h-k}, \quad 2\tan OP \cos COP = \frac{2l-h-k}{h+k+l}\tan D,$$

$$(\tan OP)^2 = \frac{(k-l)^2+(l-h)^2+(h-k)^2}{2(h+k+l)^2}(\tan D)^2.$$

99. The great circle OZ divides the triangle XOY into two right-angled triangles, and bisects the arc XY. In one of these triangles, OX is the side opposite to the right angle, one side is $\frac{1}{2}XY$, and the opposite angle is 60°. Therefore

$$\sin \tfrac{1}{2}XY = \sin OX \sin 60°.$$

But $\tan D = 2\cot OX$. Therefore the arc D depends upon XY, and may, consequently, be taken for the element of a crystal belonging to the rhombohedral system.

100. Let O be a pole of the form 1 1 1, A any pole of the form 1 0 0, M, N any poles of the forms 2 $\bar{1}$ $\bar{1}$, 1 0 $\bar{1}$ respectively. The expressions in (98) show that OM, ON are quadrants, that AOM is a multiple of 60°, and that AON is an odd multiple of 30°. Hence, the poles of the form 2 $\bar{1}$ $\bar{1}$ lie in one zone-circle, and divide it into six equal arcs; and the poles

of the form $1\,0\,\bar{1}$ bisect the arcs joining the adjacent poles of the form $2\,\bar{1}\,\bar{1}$. The poles of the form $2\,1\,\bar{1}$ are in the zone-circles containing the poles of the form $1\,1\,1$, and those of the form $1\,0\,0$. Each pair of opposite poles of the form $1\,0\,\bar{1}$ is in a zone-circle containing four poles of the form $1\,0\,0$.

101. Let O, P, Q be the poles $1\,1\,1$, $h\,k\,l$, $p\,q\,r$ respectively; and let the indices of P, Q be connected by the equations

$p = -h+2k+2l$, $q = 2h-k+2l$, $r = 2h+2k-l$. Then (98),

$$\tan AOQ = \frac{(q-r)\sqrt{3}}{2p-q-r} = \frac{(l-k)\sqrt{3}}{2h-k-l} = -\tan AOP,$$

$$\text{and} \quad 2\tan OQ \cos AOQ = \frac{2p-q-r}{p+q+r}\tan D$$

$$= \frac{k+l-2h}{h+k+l}\tan D = -2\tan OP \cos AOP.$$

Hence, $OQ = OP$, and $AOQ = 180° + AOP$. Therefore the arc PQ is bisected in O. The forms $h\,k\,l$, $p\,q\,r$ are said to be inverse with respect to each other. A combination of these two forms is called dirhombohedral. It may be denoted by $\delta\,h\,k\,l$, where $h\,k\,l$ is the symbol of any face of either of the two forms.

102. It appears from the expression for $\tan OP$, that the arcs joining the poles of the form $h\,k\,l$, and the nearest pole of the form $1\,1\,1$, are all equal. By interchanging the indices h, k, l, and changing their signs, in the expressions for $\tan AOP$, $\tan BOP$, $\tan COP$, it will be seen that the angles subtended at $1\,1\,1$ by the arcs joining any pole of the form $h\,k\,l$, and the nearest pole of the form $1\,0\,0$, are all equal. Hence, the poles of the form $h\,k\,l$ are symmetrically situated with respect to each of the three zone-circles containing the poles of the form $1\,1\,1$, and those of the form $2\,\bar{1}\,\bar{1}$. The poles of a hemihedral form with inclined faces are symmetrically situated with respect to the same zone-circles.

The poles of a dirhombohedral combination of any two holohedral forms are symmetrically situated with respect to each of seven zone-circles, six of which contain the poles of the form $1\,1\,1$, and those of the forms $2\,\bar{1}\,\bar{1}$ and $1\,0\,\bar{1}$, and the seventh contains the poles of the form $1\,0\,\bar{1}$. The poles of a dirhombohedral combination of any two hemihedral forms with inclined faces, are symmetrically situated with respect to each of the six zone-circles containing the poles of the form $1\,1\,1$, and those of the forms $2\,\bar{1}\,1$, $1\,0\,\bar{1}$. The poles of a dirhombohedral combination of any two hemihedral forms with parallel faces, are symmetrically situated with respect to the zone-circle containing the poles of the form $1\,0\,\bar{1}$.

103. The annexed figure represents the arrangement of the poles of the form $h\,k\,l$ on the surface of the sphere of projection, h being the greatest, and l algebraically the least, of three unequal indices.

If the surface of the sphere be divided into two parts by the zone-circle containing the poles of the form $1\,0\,\bar{1}$, the poles in either hemisphere will be those of a hemihedral form with inclined faces. When the algebraic sum of the indices of a form is zero, the poles of the form $h\,k\,l$ lie in the zone-circle containing the poles of the form $1\,0\,\bar{1}$. The poles in three alternate arcs joining the poles of the form $1\,0\,\bar{1}$, will be those of a hemihedral form with inclined faces.

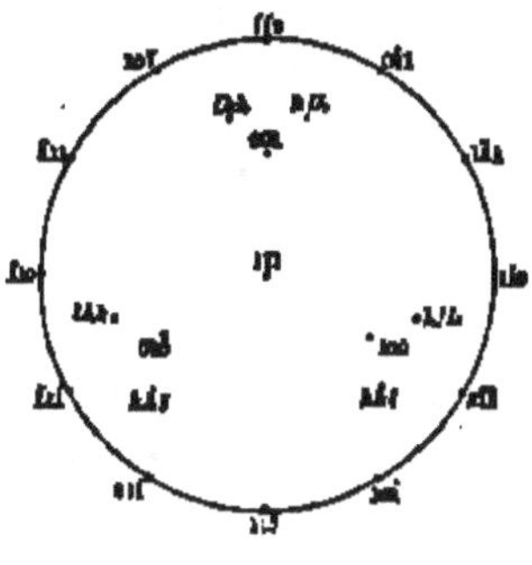

The alternate poles of the form $h\,k\,l$ are those of a hemihedral form with parallel faces.

If the surface of the sphere of projection be divided into six lunes by zone-circles through the poles of the form $1\,1\,1$, and those of the form $2\,\bar{1}\,\bar{1}$, the poles of a hemihedral form with asymmetric faces will be found in three alternate lunes.

104. The two hemihedral forms, either with inclined or with parallel faces, derived from the same holohedral form, differ only in position; for, by turning the sphere of projection through two right angles round a diameter joining any two opposite poles of the form $1\,0\,\bar{1}$, the poles of one of the hemihedral forms will change places with those of the other. The two hemihedral forms with asymmetric faces are essentially different.

105. The form $1\,1\,1$ has the two parallel faces $1\,1\,1$, $\bar{1}\,\bar{1}\,\bar{1}$. A normal to these faces is sometimes called the axis of the rhombohedron. It appears from (97) that the angles it makes with the three crystallographic axes are all equal.

106. The forms $\kappa\,1\,1\,1$, $\kappa\,\bar{1}\,\bar{1}\,\bar{1}$ consist of the faces $1\,1\,1$, $\bar{1}\,\bar{1}\,\bar{1}$ respectively.

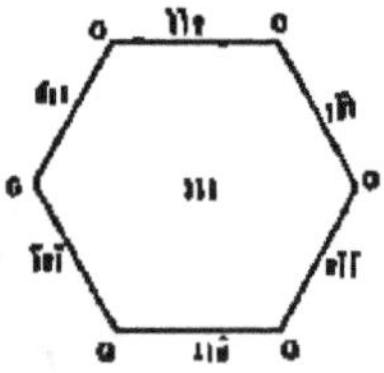

107. The form $2\,\bar{1}\,\bar{1}$ has six faces in one zone. Let G be the arc joining any two adjacent poles. Then (100) $G = 60°$.

108. Each of the forms $\kappa\,2\,\bar{1}\,\bar{1}$, $\kappa\,\bar{2}\,1\,1$ consists of three alternate faces of the form $2\,\bar{1}\,\bar{1}$.

109. The form $1\,0\,\bar{1}$ has six faces in one zone. Let H be the arc joining any two adjacent poles of the form $1\,0\,\bar{1}$. Then (100), $H = 60°$.

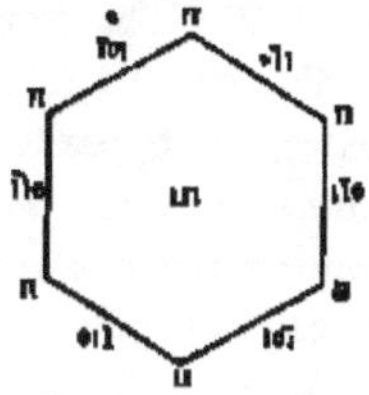

In a combination of the forms $2\bar{1}\bar{1}$, $10\bar{1}$, all the faces are in one zone the symbol of which is 111, and any face of one form makes angles of 30° with the adjacent faces of the other form. In a combination of the form 111 with the forms $2\bar{1}\bar{1}$, $10\bar{1}$, it appears from (98) that the faces of the form 111 make right angles with those of the two latter forms.

110. The form hkk, called a rhombohedron, has six faces. Let O be either pole of the form 111; A, P any two adjacent poles of the forms 100, hkk respectively; $OA = D$, $OP = T$. Let V be the arc joining any two poles of the form hkk, on the same side of the zone-circle 111; W the arc joining any two adjacent poles on opposite sides of the zone-circle 111. The poles of the form hkk are in the zone-circle containing the poles of the forms 111 and 100, therefore (98) the arc V subtends an angle of 120° at O. Hence, making $l = k$ in the expression for $(\tan OP)^2$, we have

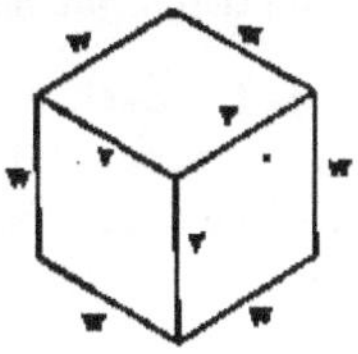

$$\tan T = \frac{h-k}{h+2k} \tan D, \quad \sin \tfrac{1}{2} V = \sin 60° \sin T, \quad W = 180° - V.$$

The position of a rhombohedron is said to be direct or inverse according as $\tan T$ is positive or negative, or, according as OP, OA are measured in the same or in opposite directions from O.

In a combination of the forms $10\bar{1}$, hkk, each face of the form 101 is in a zone containing four faces of the form hkk. The arcs joining any pole of the form hkk and the poles of the form $10\bar{1}$, are $90° - \frac{1}{2}V$, $90°$, $90° + \frac{1}{2}V$.

111. Each of the forms κhkl, $\kappa \bar{h}\bar{k}\bar{l}$ consists of three faces of the form hkk, making equal angles with one another.

112. The form $h\,k\,l$, where $h+k+l=0$, has twelve faces in the zone $1\,1\,1$. Let H be the arc joining any two adjacent poles, on opposite sides of a pole of the form $2\,\bar{1}\,\bar{1}$, h being numerically the largest index; W the arc joining any two adjacent poles, on opposite sides of a pole of the form $1\,0\,\bar{1}$. Then (98), since $h+k+l=0$, the arc joining the pole $1\,1\,1$, and any pole of the form $h\,k\,l$, is a quadrant. Hence

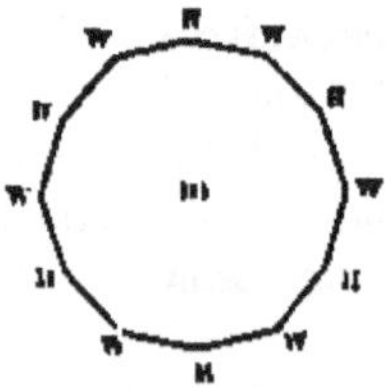

$$\tan\tfrac{1}{2}H=\frac{(k-l)\sqrt{3}}{2h-k-l},\quad W=60°-H.$$

In a combination of this form with the form $1\,1\,1$, the faces of the two forms make right angles with one another.

113. Each of the forms $\kappa\,h\,k\,l$, $\kappa\,\bar{h}\,\bar{k}\,l$, where $h+k+l=0$, has the faces of the form $h\,k\,l$, which meet in alternate edges H. The angles between any two adjacent faces are alternately H and $120°-H$.

114. Each of the forms $\pi\,h\,k\,l$, $\pi\,l\,k\,h$, where $h+k+l=0$, consists of alternate faces of the form $h\,k\,l$. The angle between any two adjacent faces is $60°$.

115. Each of the forms $\alpha\,h\,k\,l$, $\alpha\,l\,k\,h$, where $h+k+l=0$, consists of the faces of the form $h\,k\,l$, which meet in the alternate edges W. The angles between any two adjacent faces are alternately W and $120°-W$.

116. The form $h\,k\,l$ has twelve faces. Let D, T be the arcs joining any poles of the forms $1\,0\,0$, $h\,k\,l$ respectively, and the nearest poles of the form $1\,1\,1$; H, K, L the arcs joining any two poles of the form $h\,k\,l$, equidistant from the pole $1\,1\,1$, in the symbols of which h, k, l occupy the same places; W the arc joining any two adjacent poles unequally distant from the pole $1\,1\,1$; 2θ, 2ϕ, 2ψ the angles subtended at the pole $1\,1\,1$ by the arcs H, K, L. Then (98),

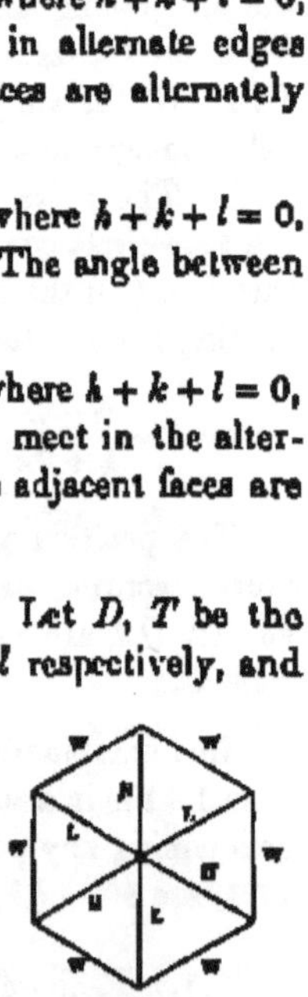

$$\tan\theta = \frac{(k-l)\sqrt{3}}{2h-k-l}, \quad \tan\phi = \frac{(l-h)\sqrt{3}}{2k-l-h}, \quad \tan\psi = \frac{(h-k)\sqrt{3}}{2l-h-k},$$

$$(\tan T)^2 = \frac{(k-l)^2 + (l-h)^2 + (h-k)^2}{2(h+k+l)^2} (\tan D)^2.$$

In the triangles having their vertex in the pole 1 1 1, and the bases H, K, L, the sides which meet in 1 1 1 are each equal to T. Hence

$$\sin \tfrac{1}{2}H = \sin\theta \sin T, \quad \sin \tfrac{1}{2}K = \sin\phi \sin T,$$

$$\sin \tfrac{1}{2}L = \sin\psi \sin T, \quad W = 180^\circ - K.$$

When $2k = h + l$, the angles H, L are equal, and the edges W are parallel to the faces of the form 1 1 1.

In a combination of the forms $1\,0\,\bar{1}$, $h\,k\,l$, each face of the form $1\,0\,\bar{1}$ is in a zone containing four faces of the form $h\,k\,l$. The arcs joining any pole of the form $h\,k\,l$ and the poles of the form $1\,0\,\bar{1}$ are $90^\circ \mp \tfrac{1}{2}H$, $90^\circ \mp \tfrac{1}{2}K$, $90^\circ \mp \tfrac{1}{2}L$.

117. Each of the forms $\kappa\,h\,k\,l$, $\kappa\,\bar{h}\,\bar{k}\,\bar{l}$, consists of six faces of the form $h\,k\,l$, the poles of which are equidistant from a pole of the form 1 1 1.

118. Each of the forms $\pi\,h\,k\,l$, $\pi\,l\,k\,h$, is contained by alternate pairs of parallel faces of the form $h\,k\,l$. Let V be the arc joining any two alternate poles of the form $h\,k\,l$, equally distant from a pole of the form 1 1 1. Then V will subtend an angle of 120° at 1 1 1. Therefore $\sin \tfrac{1}{2}V = \sin 60^\circ \sin T$.

119. Each of the forms $\alpha\,h\,k\,l$, $\alpha\,l\,k\,h$, is contained by pairs of faces of the form $h\,k\,l$, which meet in alternate edges W. The arc joining any two poles equidistant from the pole 1 1 1, is V, and the greater of the arcs joining two adjacent poles unequally distant from 1 1 1, is $180^\circ - H$.

120. The principal cleavages are parallel to the faces of one of the forms 1 1 1, $1\,0\,\bar{1}$, $2\,\bar{1}\,\bar{1}$, $h\,k\,k$.

121. Let P, Q, R be three poles of a rhombohedron, equidistant from O, the pole 1 1 1; and let the zone-circle through P, Q, contain S, a pole of another rhombohedron. S is in the zone-circle OR which bisects the angle POQ and the arc PQ. The angle $POQ = 120°$, and therefore $SOP = 60°$, $OSP = 90°$, and $\cos SOP = \tan OS \cot OP$, $\cos 60° = \frac{1}{2}$, therefore

$$\tan OP = 2 \tan OS.$$

122. Let O, A be the poles 1 1 1, 1 0 0 respectively; P, Q any two poles the symbols of which are known. Then (98) $\tan AOP$, $\tan AOQ$ can be found in terms of the indices of P and Q, therefore $\tan POQ$ is known in terms of the same indices; also $\tan OP$, $\tan OQ$ can be expressed in terms of $\tan D$ and the indices of P and Q. Therefore, knowing OP, OQ, two sides of a spherical triangle, and POQ the included angle, the third side PQ may be found.

123. Let V be the arc joining any two of three equidistant poles of the form $h\,k\,k$. Then (110),

$$\sin \tfrac{1}{2} V = \sin 60° \sin T, \quad \frac{h-k}{h+2k} = \frac{\tan T}{\tan D},$$

$\tan T$ being positive or negative according as T and D are measured from the pole 1 1 1 in the same or in opposite directions. Hence, when D and V are known, the ratio of h to k may be found.

124. Let H be the arc joining any two poles of the form $h\,k\,l$, where $h+k+l=0$, in which the largest index holds the same place. Then, if the arc, not being a multiple of 60°, which joins any two poles of the form, be given, we can find H. The ratios of the indices can then be found by means of the equations

$$\tan \tfrac{1}{2} H = \frac{(k-l)\sqrt{3}}{2h-k-l}, \quad h+k+l=0.$$

125. Suppose the arcs joining any pole of the form $h\,k\,l$, and each of two other poles of the same form, the three poles not being in the same zone-circle, and the arc D, to be given. The given arcs or their supplements will be two of the arcs H, K, L, V (116), (118). By eliminating T between the equations in (116), (118), observing that $\phi - \theta = 60°$, $\psi + \theta = 60°$, we obtain

$$\frac{\tan\theta}{\tan 60°} = \frac{\tan\frac{1}{2}(K-L)}{\tan\frac{1}{2}(K+L)}, \quad \frac{\sin\theta}{\sin 60°} = \frac{\sin\frac{1}{2}H}{\sin\frac{1}{2}V},$$

$$\frac{\tan\phi}{\tan 60°} = \frac{\tan\frac{1}{2}(L+H)}{\tan\frac{1}{2}(L-H)}, \quad \frac{\sin\phi}{\sin 60°} = \frac{\sin\frac{1}{2}K}{\sin\frac{1}{2}V},$$

$$\frac{\tan\psi}{\tan 60°} = \frac{\tan\frac{1}{2}(K-H)}{\tan\frac{1}{2}(K+H)}, \quad \frac{\sin\psi}{\sin 60°} = \frac{\sin\frac{1}{2}L}{\sin\frac{1}{2}V}.$$

Two of the arcs H, K, L, V, and D, being known, T and one of the angles θ, ϕ, ψ may be found. The ratios of the indices may then be obtained from two of the equations

$$\tan\theta = \frac{(k-l)\sqrt{3}}{2h-k-l}, \quad 2\tan T\cos\theta = \frac{2h-k-l}{h+k+l}\tan D,$$

$$\tan\phi = \frac{(l-h)\sqrt{3}}{2k-l-h}, \quad 2\tan T\cos\phi = \frac{2k-l-h}{h+k+l}\tan D,$$

$$\tan\psi = \frac{(h-k)\sqrt{3}}{2l-h-k}, \quad 2\tan T\cos\psi = \frac{2l-h-k}{h+k+l}\tan D.$$

126. When the arc joining two poles of either of the forms $h\,k\,k$, $h\,k\,l$, and the symbols of the poles, are known, the expressions in (110) or (116) enable us to find the angle which the given arc subtends at the pole 1 1 1, and T, the arc joining either pole and the pole 1 1 1. Then, knowing $\tan T$ and the indices of the form, $\tan D$ may be found.

127. Let O be the pole 1 1 1; R, S any two poles in a zone-circle passing through O; p q r the symbol of any zone-circle passing through R, except RS; $u\,v\,w$ the symbol of S; and suppose the arc RS to be given. Let P be the intersection

of the zone-circle RS and the zone-circle $\bar{1}\,1\,1$, PS being greater than PR. Then $1\,1\,1$ is the symbol of a zone-circle passing through P; the symbol of O is $1\,1\,1$; and OP is a quadrant. Therefore (20),

$$\sin(2PR + RS) = (2i - 1)\sin RS,$$

where
$$i = \frac{u+v+w}{3}\,\frac{p+q+r}{pu+qv+rw}.$$

Having found OR or OS by means of the preceding equation, $\tan D$ is given by (98).

128. Let O, A be the poles $1\,1\,1$, $1\,0\,0$ respectively; R, S any two poles; $p\,q\,r$ the symbol of any zone-circle containing R, except RS; $u\,v\,w$ the symbol of S, and suppose the arc RS to be given. Let P be the intersection of RS and the zone-circle $1\,1\,1$, PS being greater than PR; Q the intersection of RS and a zone-circle having for its symbol the symbol of P, and therefore passing through O, for the symbol of a pole in the zone-circle $1\,1\,1$ is the symbol of a zone-circle containing the pole $1\,1\,1$. It is easily proved that $\tan AOQ = -\cot AOP$. Hence POQ is a right angle, and PQ is a quadrant. Let $h\,k\,l$ be the symbol of Q, the indices of Q being deduced from those of R and S. Then (20),

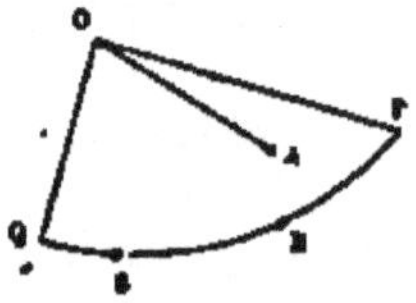

$$\sin(2PR + RS) = (2i - 1)\sin RS,$$

where
$$i = \frac{u+v+w}{h+k+l}\,\frac{ph+qk+rl}{pu+qv+rw}.$$

Having found PR or PS by means of the preceding equation, and $\tan POR$ or $\tan POS$, we have

$$\cos PR = \cos POR \sin OR, \quad \cos PS = \cos POS \sin OS.$$

Hence, knowing OR or OS, $\tan D$ is given by (98).

CHAPTER V.

PRISMATIC SYSTEM.

129. In the prismatic system the axes make right angles with one another.

130. The form $h\,k\,l$ consists of the faces in the symbols of which each of the indices h, k, l may be either positive or negative, but always occupies the same place. When h, k, l are all finite, the form has the eight faces

$$\begin{array}{cccc} h\,k\,l & h\,\bar{k}\,\bar{l} & \bar{h}\,k\,\bar{l} & \bar{h}\,\bar{k}\,l \\ \bar{h}\,\bar{k}\,\bar{l} & \bar{h}\,k\,l & h\,\bar{k}\,l & h\,k\,\bar{l} \end{array}$$

When one of the indices is zero, the number of faces will be four. When two of the indices are zero, the number of faces will be two.

131. The form contained by the faces of the form $h\,k\,l$, which have an odd number of positive indices, or by the faces of the form $h\,k\,l$, which have an odd number of negative indices, is said to be hemihedral with asymmetric faces, and will be denoted by the symbol $\alpha\,h\,k\,l$, where $h\,k\,l$ is the symbol of any one of its faces. The upper and lower lines of the table in (130) contain the symbols of the two half forms respectively.

132. The form consisting of the faces of the form $h\,k\,l$, in the symbols of which the sign of one of the indices remains unchanged, is said to be hemihedral with inclined faces, and

may be denoted by the symbol $\kappa\, h\, k\, l$, the index which preserves its sign unchanged having that sign either prefixed or placed over it.

133. The form having the faces of $h\, k\, l$, in which two of the indices change their signs together, is said to be hemihedral with parallel faces, and may be denoted by the symbol $\pi\, \dot{h}\, k\, l$, a dot being placed over the index the sign of which is independent of the signs of the other two indices.

134. Let a, b, c be the parameters; A, B, C the poles $1\,0\,0$, $0\,1\,0$, $0\,0\,1$ respectively; P the pole $h\, k\, l$. The axes make right angles with one another, therefore the sides of the triangle

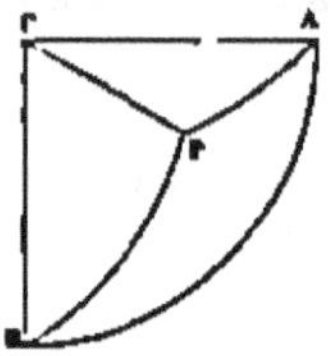

XYZ are quadrants, its angles are right angles, and X, Y, Z are the poles of YZ, ZX, XY. But A, B, C are the poles of YZ, ZX, XY, and they have no negative indices, therefore (3) A, B, C coincide with X, Y, Z respectively. Hence, the sides of the triangle ABC are quadrants, and its angles are right angles. The quadrantal triangles PBC, PCA, PAB give

$$\cos AP = \sin BP \cos ABP = \sin CP \cos ACP,$$
$$\cos BP = \sin CP \cos BCP = \sin AP \cos BAP,$$
$$\cos CP = \sin AP \cos CAP = \sin BP \cos CBP.$$

$$\cot AP = \tan BCP \cos BAP = \tan CBP \cos CAP,$$
$$\cot BP = \tan CAP \cos CBP = \tan ACP \cos ABP,$$
$$\cot CP = \tan ABP \cos ACP = \tan BAP \cos BCP.$$

Also, since A, B, C coincide with X, Y, Z,

$$\frac{a}{h} \cos AP = \frac{b}{k} \cos BP = \frac{c}{l} \cos CP.$$

Hence, substituting in the preceding equations the values of cos AP, cos BP, cos CP given above, and observing that

$$\cos CAP = \sin BAP,\ \cos ABP = \sin CBP,\ \cos BCP = \sin ACP,$$

we obtain

$$\tan BAP = \frac{l}{k}\frac{b}{c},\quad \tan CBP = \frac{h}{l}\frac{c}{a},\quad \tan ACP = \frac{k}{h}\frac{a}{b}.$$

135. Let D be the arc joining the poles 0 1 0, 0 1 1; E the arc joining the poles 0 0 1, 1 0 1; F the arc joining the poles 1 0 0, 1 1 0. Then, since the sides of the triangle ABC are quadrants, the arcs D, E, F measure the angles they respectively subtend at A, B, C. Therefore

$$\tan D = \frac{b}{c},\quad \tan E = \frac{c}{a},\quad \tan F = \frac{a}{b}.$$

Hence

$$\tan BAP = \frac{l}{k}\tan D,\quad \tan CBP = \frac{h}{l}\tan E,\quad \tan ACP = \frac{k}{h}\tan F,$$

$$\cot AP = \frac{h}{k}\cot F\cos BAP = \frac{h}{l}\tan E\cos CAP,$$

$$\cot BP = \frac{k}{l}\cot D\cos CBP = \frac{k}{h}\tan F\cos ABP,$$

$$\cot CP = \frac{l}{h}\cot E\cos ACP = \frac{l}{k}\tan D\cos BCP.$$

136. Since the ratios of the parameters can be expressed in terms of the tangents of any two of the arcs D, E, F, and their product, any two of the arcs D, E, F may be taken for the elements of the crystal. The arcs D, E, F are connected by the equation

$$\tan D\tan E\tan F = 1.$$

137. It appears from (135) that the arcs joining either pole of the form 1 0 0, and the adjacent poles of the form $h\,k\,l$, are all equal; that the arcs joining either pole of the form 0 1 0,

and the adjacent poles of the form $h\,k\,l$, are all equal; and that the arcs joining either pole of the form 0 0 1, and the adjacent poles of the form $h\,k\,l$, are all equal. Hence, the poles of the form $h\,k\,l$ are symmetrically arranged with respect to each of the zone-circles 1 0 0, 0 1 0, 0 0 1.

The poles of a hemihedral form with inclined faces are symmetrically arranged with respect to two of the zone-circles 1 0 0, 0 1 0, 0 0 1, the first, second or third being excluded, according as the first, second or third index preserves its sign unchanged.

The poles of a hemihedral form with parallel faces are symmetrically situated with respect to the zone-circles 1 0 0, 0 1 0, 0 0 1, according as the sign of the first, second or third index is independent of the signs of the other two indices.

139. The annexed figure represents the arrangement of the poles of the forms $h\,k\,l$, $0\,k\,l$, $h\,0\,l$, $h\,k\,0$, 1 0 0, 0 1 0, 0 0 1 on the surface of the sphere of projection.

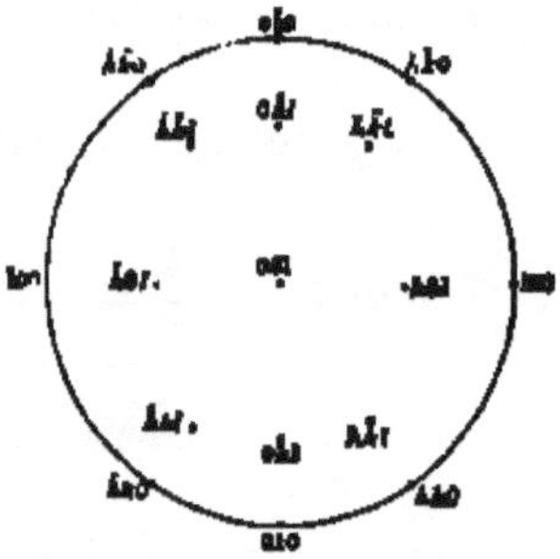

A hemihedral form with asymmetric faces has the alternate poles of the form $h\,k\,l$.

The poles of a hemihedral form with inclined faces are contained in one of the two hemispheres, into which the sphere of projection is divided by one of the zone-circles 1 0 0, 0 1 0, 0 0 1.

If the surface of the sphere be divided into four lunes by two of the zone-circles 1 0 0, 0 1 0, 0 0 1, the poles of a hemihedral form with parallel faces will be found in two alternate lunes.

139. The two hemihedral forms with either inclined or parallel faces, derived from the same holohedral form, differ only in position; for, by making the sphere revolve through two right angles round the poles of one of the forms 1 0 0, 0 1 0, 0 0 1, the poles of one half-form will change places with those of the other. The two hemihedral forms with asymmetric faces, derived from the same holohedral form, are essentially different.

140. The three forms 1 0 0, 0 1 0, 0 0 1 have each two parallel faces. The arc joining poles of any two of three forms is a quadrant (134). Hence, in a combination of these forms with one another, the faces of each form make right angles with those of the other two.

Either of the preceding forms may become hemihedral.

141. The form 0 *k* *l* has four faces. Let *L* be the arc joining any two adjacent poles differing in the signs of *L*. Then $\frac{1}{2}L = 0\,1\,0, 0\,k\,l$. Hence (135),

$$\tan\tfrac{1}{2}L = \frac{l}{k}\tan D, \quad K = 180^\circ - L.$$

The arc joining either pole of the form 1 0 0, and any pole of the form 0 *k* *l*, is a quadrant. Therefore, in a combination of the forms 1 0 0, 0 *k* *l*, the faces of the two forms make right angles with one another.

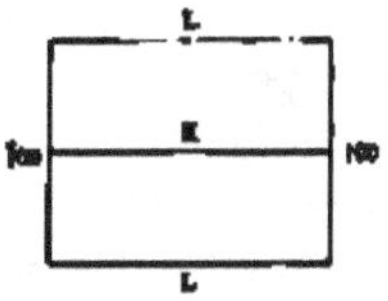

142. The form *h* 0 *l* has four faces. Let *H* be the arc joining any two adjacent poles differing in the signs of *h*. Then $\frac{1}{2}H = 0\,0\,1, h\,0\,l$. Hence (135),

$$\tan\tfrac{1}{2}H = \frac{h}{l}\tan E, \quad L = 180^\circ - H.$$

The arc joining either pole of the form 0 1 0, and any pole of the form $h\,0\,l$, is a quadrant. Hence, in a combination of the forms 0 1 0, $h\,0\,l$, the faces of the two forms make right angles with one another.

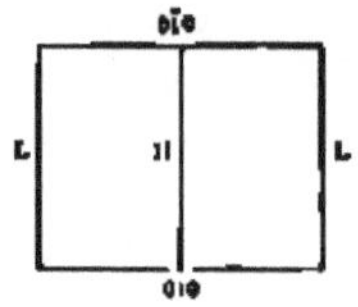

143. The form $h\,k\,0$ has four faces. Let K be the arc joining any two adjacent poles differing in the signs of k. Then $\frac{1}{2}K = 1\,0\,0, h\,k\,0$. Hence (135)

$$\tan \tfrac{1}{2}K = \frac{k}{h} \tan F, \quad H = 180^\circ - K.$$

The arc joining either pole of the form 0 0 1, and any pole of the form $h\,k\,0$, is a quadrant. Hence, in a combination of the forms 0 0 1, $h\,k\,0$, the faces of the two forms make right angles with one another.

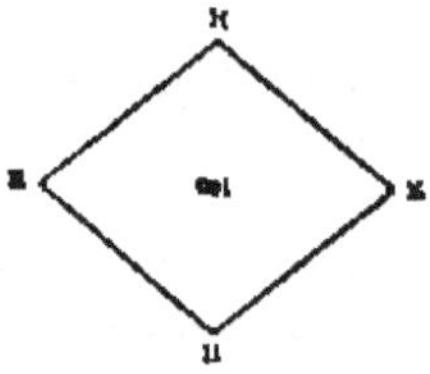

144. When either of the forms $0\,k\,l$, $h\,0\,l$, $h\,k\,0$ becomes hemihedral with inclined faces, the hemihedral form consists of two adjacent faces.

When either of them becomes hemihedral with parallel faces, the hemihedral form consists of two opposite faces.

145. The form $h\,k\,l$ has eight faces. Let H, K, L be the arcs joining any two adjacent poles differing in the signs of h, k, l respectively. Then $90^\circ - \frac{1}{2}H = 1\,0\,0, h\,k\,l$; $90^\circ - \frac{1}{2}K = 0\,1\,0, h\,k\,l$; $90^\circ - \frac{1}{2}L = 0\,0\,1, h\,k\,l$. Hence (135), (134), ϕ being the angle which the arc $1\,0\,0, h\,k\,l$ subtends at 0 0 1,

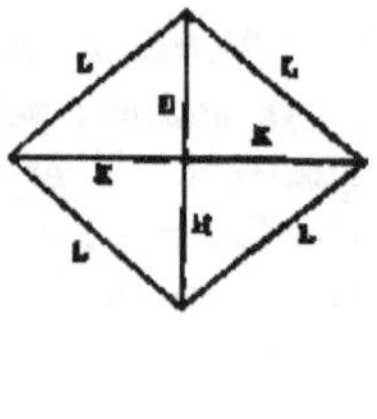

$$\tan \phi = \frac{k}{h} \tan F, \quad \tan \tfrac{1}{2}L = \frac{l}{h} \cot E \cos \phi,$$

$$\sin \tfrac{1}{2}K = \cos \tfrac{1}{2}L \sin \phi, \quad \sin \tfrac{1}{2}H = \cos \tfrac{1}{2}L \cos \phi.$$

146. A hemihedral form with asymmetric faces is a four sided figure contained by the alternate faces of the form $h\,k\,l$.

147. A hemihedral form with inclined faces consists of four faces making one of the solid angles of the form $h\,k\,l$.

148. A hemihedral form with parallel faces has four faces of the form $h\,k\,l$ in one zone.

149. Let A, B, C be the poles 1 0 0, 0 1 0, 0 0 1 respectively; P the pole $h\,k\,l$; Q the pole $p\,q\,r$. Then as in (84), when Q is in the zone-circle AP,

$$\frac{h}{p}\frac{\tan AP}{\tan AQ}=\frac{k}{q}=\frac{l}{r}.$$

When Q is in the zone-circle BP,

$$\frac{k}{q}\frac{\tan BP}{\tan BQ}=\frac{l}{r}=\frac{h}{p}.$$

When Q is in the zone-circle CP,

$$\frac{l}{r}\frac{\tan CP}{\tan CQ}=\frac{h}{p}=\frac{k}{q}.$$

150. Let U, V be any two of the three poles 1 0 0, 0 1 0, 0 0 1; P, Q any two poles the symbols of which are given. Then, knowing two of the arcs D, E, F, and the symbols of P, Q, we can find UP, UQ, VUP, VUQ by (135). Hence knowing UP, UQ and PUQ, the arc PQ can be found.

151. If the arc joining any two poles, not opposite to one another, of one of the forms $0\,k\,l$, $h\,0\,l$, $h\,k\,0$, be given, the ratio of the indices may be obtained from (141), (142) or (143).

152. In the form $h\,k\,l$, the arcs joining any pole, and each of two others, no two of the poles being opposite to one another, or their supplements, will be two of the arcs H, K, L. Therefore two of the arcs H, K, L being known, we can find ϕ, and thence the ratios of h, k, l, by (145).

153. The arcs D, E, F may be found from the expressions in (141), (142) or (143), having given the arcs joining any two

poles, not opposite to one another, of any two of the forms $0\,k\,l$, $h\,0\,l$, $h\,k\,0$; or, from the expressions in (145), having given the arcs joining any pole of the form $h\,k\,l$, and each of two other poles, the three poles not being in the same zone-circle.

154. Let U, V be any two of the three poles 1 0 0, 0 1 0, 0 0 1; P, Q any two poles the symbols of which are known; and suppose the arcs UP, VQ to be given. Then, T being the intersection of the zone-circles UP, VQ, the symbol of T is known by (5), (7); and the arcs UT, VT by (149). The quadrantal triangle UTV gives the angles UVT, VUT, whence the arcs D, E, F may be found by (135).

155. Let U, V be any two of the three poles 1 0 0, 0 1 0, 0 0 1; P, R two poles in a zone-circle containing U; Q, S two poles in a zone-circle containing V. Two of the zone-circles containing every two of the poles 1 0 0, 0 1 0, 0 0 1, will have U, V respectively for poles. Let PR, QS meet these zone-circles in M, N respectively. Then UM, VN will be quadrants. Hence, if the arcs PR, QS, and the symbols of P, R, Q, S be given, the arcs UP, VQ become known by (20), and then the arcs D, E, F may be found by (154).

156. The arcs D, E, F may also be found from the arcs joining three given poles in one zone-circle not passing through any one of the poles 1 0 0, 0 1 0, 0 0 1. Let P, Q, R be the given poles; A, B, C the poles 1 0 0, 0 1 0, 0 0 1 respectively. Let L, L' be the intersections of PR and BC; M, M' those of PR and CA; N, N' those of PR and AB; and let x be the less of the arcs NM, MN'; y the less of the arcs NL, LN'; z the less of the arcs ML, LM'. Then, knowing the symbols of P, Q, R, and the arcs joining P, Q, R, the symbols of L, M, N may be found by (5), (7), and the arcs PL, PM, PN by (13) or

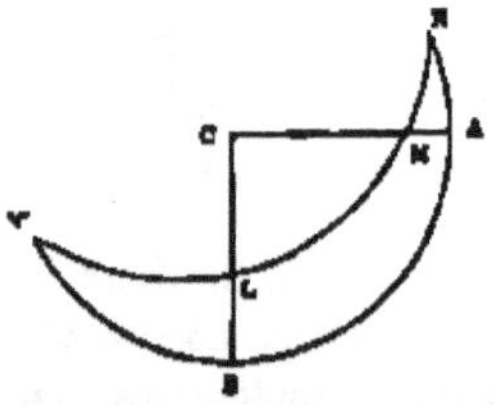

(14). Hence the arcs joining L, M, N are known. It is easily seen that

$$\frac{\tan BL}{\tan CL} = \frac{\tan LN'}{\tan LM'}, \quad \frac{\tan CM}{\tan AM} = \frac{\tan LM}{\tan MN'}, \quad \frac{\tan AN}{\tan BN'} = \frac{\tan MN}{\tan LN'},$$

and that

$$\tan CL = \cot BL, \quad \tan AM = \cot CM, \quad \tan BN' = \cot AN.$$

Hence
$$(\tan BL)^2 = \tan y \cot z,$$
$$(\tan CM)^2 = \tan z \cot x,$$
$$(\tan AN)^2 = \tan x \cot y.$$

Then, knowing $\tan BL$, $\tan CM$, $\tan AN$, and the symbols of L, M, N, the arcs D, E, F are given by (135).

CHAPTER VI.

OBLIQUE SYSTEM.

157. In the oblique system one axis (OY) makes right angles with each of the other two axes.

158. The form $h\,k\,l$ consists of the faces in the symbols of which $\pm h$, $\pm k$, $\pm l$ occupy the same places respectively, and h and l change their signs together. When k is finite, the form has the four faces

$$h\,k\,l \qquad \bar{h}\,k\,\bar{l} \qquad h\,\bar{k}\,l \qquad \bar{h}\,\bar{k}\,\bar{l}$$

When k is zero, or when the symbol of the form is 0 1 0, the number of faces will be two.

159. The hemihedral form has the faces of the form $h\,k\,l$ in the symbols of which the sign of k does not change. It may be denoted by $\kappa\,h\,k\,l$, where $h\,k\,l$ is the symbol of either of its faces. The poles of the two half forms are on opposite sides of the zone-circle 0 1 0.

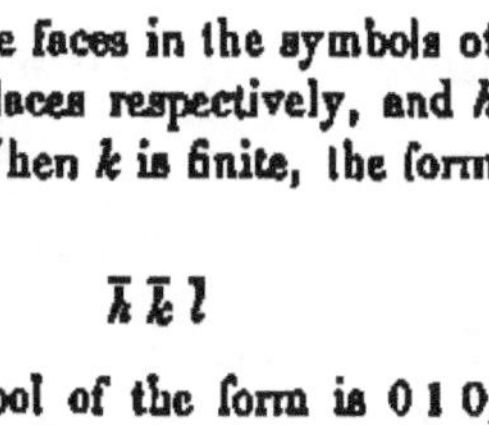

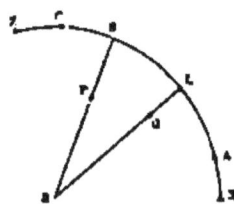

160. Let a, b, c be the parameters; A, B, C the poles 1 0 0, 0 1 0, 0 0 1; G the pole 1 1 1; P the pole $h\,k\,l$. The axis OY makes right angles with each of the other two axes, therefore YZ, YX are quadrants. But YA, ZA, ZB, XB, XC,

YC are quadrants (3). Hence, B coincides with Y; the poles C, A are in the great circle ZX; and BC, BA are quadrants. Let the zone-circle CA meet the zone-circle BP in S, and the zone-circle BG in L. The symbols of S, L will, therefore, be $h\,0\,l$ and $1\,0\,1$ respectively.

But $$\frac{a}{h}\cos XP = \frac{b}{k}\cos BP = \frac{c}{l}\cos ZP,$$

$\cos XP = \sin BP \sin CS$, and $\cos ZP = \sin BP \sin AS$. Therefore

$$\frac{a}{h}\sin CS = \frac{b}{k}\cot BP = \frac{c}{l}\sin AS.$$

Hence $$a \sin CL = b \cot BG = c \sin AL.$$

These equations give

$$\frac{\sin CL}{\sin AL}\,\frac{\sin AS}{\sin CS} = \frac{l}{h}.$$ Therefore, putting

$$\tan\theta = \frac{h}{l}\,\frac{\sin CL}{\sin AL}, \text{ and consequently } \frac{\sin CS}{\sin AS} = \tan\theta,$$

we obtain $$\tan(AS - \tfrac{1}{2}AC) = \tan\tfrac{1}{2}AC \tan(\tfrac{1}{4}\pi - \theta).$$

Also $$\frac{\tan BP}{\tan BG} = \frac{h}{k}\,\frac{\sin CL}{\sin CS} = \frac{l}{k}\,\frac{\sin AL}{\sin AS}.$$

$$\cos AP = \sin BP \cos AS, \quad \cos CP = \sin BP \cos CS.$$

161. The arc ZX is the supplement of AC, or of $AL + CL$, and the ratios of the parameters are given in terms of $\sin AL$, $\cot BG$, $\sin CL$. Hence the arcs AL, BG, CL may be taken for the elements of a crystal of the oblique system.

162. The arc joining two poles of the form $h\,k\,l$ differing only in the signs of k, is manifestly bisected at right angles by the zone-circle $0\,1\,0$. The poles of the form $h\,k\,l$ are, therefore, symmetrically situated with respect to the zone-circle $0\,1\,0$.

The arc joining the two poles of the form $\kappa\,h\,k\,l$ is bisected in a pole of the form $0\,1\,0$.

163. The form 0 1 0 has two parallel faces.

164. The form $h\,0\,l$ has two parallel faces in the zone 0 1 0. Let S be the pole $h\,0\,l$. Then (160) BS is a quadrant; the arc AS is given by the equations

$$\tan\theta = \frac{h}{l}\,\frac{\sin CL}{\sin AL},\quad \tan(AS - \tfrac{1}{2}AC) = \tan\tfrac{1}{2}AC\tan(\tfrac{1}{4}\pi - \theta);$$

and CS is either the difference or sum of AC and AS.

165. The form $h\,k\,l$ has four faces. Their poles are in a zone-circle passing through the poles of the form 0 1 0. Let K be the arc joining any two adjacent poles differing in the signs of k; P the pole $h\,k\,l$. Then $K = 180° - 2BP$, where BP is given by the equations

$$\tan\theta = \frac{h}{l}\,\frac{\sin CL}{\sin AL},\quad \tan(AS - \tfrac{1}{2}AC) = \tan\tfrac{1}{2}AC\tan(\tfrac{1}{4}\pi - \theta),$$

$$\frac{\tan BP}{\tan BG} = \frac{h}{k}\,\frac{\sin CL}{\sin CS} = \frac{l}{k}\,\frac{\sin AL}{\sin AS}.$$

The arcs AP, CP are given by the equations

$$\cos AP = \sin BP\cos AS,\quad \cos CP = \sin BP\cos CS.$$

166. The form $\kappa\,h\,k\,l$ has two faces of the form $h\,k\,l$, the poles of which are equidistant from a pole of the form 0 1 0. The arc joining the two poles is equal to $2BP$.

167. Suppose the arc AS in (164) to be given. Then the ratio of the indices of the form $h\,0\,l$ can be found from the equation

$$\frac{l}{h} = \frac{\sin AL}{\sin CL}\,\frac{\sin CS}{\sin AS}.$$

168. Suppose any two of the arcs AP, BP, CP in (165) to be given. Then, having found the arcs AS, BP, the ratios of the indices of the form $h\,k\,l$ are given by the equations

$$\frac{h}{l} = \frac{\sin AL}{\sin CL}\,\frac{\sin CS}{\sin AS},\quad \frac{k}{l} = \frac{\sin AL}{\sin AS}\,\frac{\tan BG}{\tan BP}.$$

169. Let B, P, Q be the poles 0 1 0, $h\,k\,l$, $p\,q\,r$ respectively. Then, when Q is in the zone-circle BP, it appears from the equations between $\cot BP$, $\sin AS$, $\sin CS$ in (160) that

$$\frac{k}{q}\frac{\tan BP}{\tan BQ}=\frac{l}{r}=\frac{h}{p}.$$

170. Let A, B, C be the poles 1 0 0, 0 1 0, 0 0 1 respectively; P the pole $h\,k\,l$; Q the pole $p\,q\,r$. Let BP, BQ meet CA in S, T respectively. Then BP, BQ, AS, AT may be found by (160). Hence, knowing the arcs BP, BQ, and the included angle PBQ, which is measured by ST, the difference between AS and AT, the arc PQ can be found by the rules of spherical trigonometry.

171. Let A, B, C be the poles 1 0 0, 0 1 0, 0 0 1; G, L the poles 1 1 1, 1 0 1; P the pole $h\,k\,l$. Suppose the arcs AP, BP, CP to be given. Let BP meet CA in S. Then (160),

$$\cos CP=\sin BP\cos CS,\quad \cos AP=\sin BP\cos AS,$$

whence CS, AS, AC are known. But

$$\frac{\sin CL}{\sin AL}=\frac{l}{h}\frac{\sin CS}{\sin AS}.\ \text{Hence, putting}\ \tan\theta=\frac{l}{h}\frac{\sin CS}{\sin AS},$$

$$\tan(AL-\tfrac{1}{2}AC)=\tan\tfrac{1}{2}AC\tan(\tfrac{1}{4}\pi-\theta).$$

Having found AL, CL by means of the preceding equations, BG is given by

$$\frac{\tan BG}{\tan BP}=\frac{k}{h}\frac{\sin CS}{\sin CL}=\frac{k}{l}\frac{\sin AS}{\sin AL}.$$

Hence AL, BG, CL, the angular elements of the crystal, are known.

172. Let A, B, C be the poles 1 0 0, 0 1 0, 0 0 1; P any pole not in CA; U, V, W three poles in CA; and suppose the arcs BP, UV, VW, and the symbols of P, U, V, W to be given. Let BP meet CA in S. The symbol of S may be found by

(5) and (7); and the arcs CU, AU, SU by (13) or (14). Therefore, knowing AS, BP, CS, the elements of the crystal may be found by (171).

Let Q be a pole in a zone-circle BP, and suppose that the arc PQ, and the symbol of Q had been given, instead of the arc BP. Then, since BS is a quadrant, the arc BP may be found by (20), and the elements of the crystal by the method given above.

173. Let A, B, C be the poles 1 0 0, 0 1 0, 0 0 1; P, Q any two poles of different forms, not in CA; and suppose the

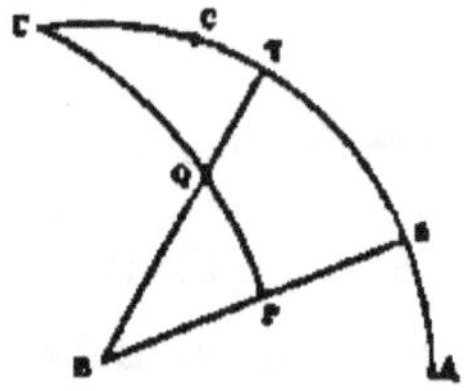

arcs BP, BQ, PQ, and the symbols of P and Q, to be given. Let BP, BQ, PQ meet CA in S, T, U respectively. The symbols of S, T, U can be found from those of P and Q; and the arcs SU, TU can be computed from BP, BQ, PQ. The arcs AS, CS are then given by (13) or (14); and the elements of the crystal may be found by (171) from AS, BP, CS, or from AT, BQ, CT.

CHAPTER VII.

ANORTHIC SYSTEM.

174. In the anorthic system the form $h\,k\,l$ has the two parallel faces $h\,k\,l$, $\bar{h}\,\bar{k}\,\bar{l}$.

175. Let a, b, c be the parameters; A, B, C the poles $1\,0\,0$, $0\,1\,0$, $0\,0\,1$; G the pole $1\,1\,1$; D, E, F the points in which AG, BG, CG intersect BC, CA, AB, and, therefore, the poles $0\,1\,1$, $1\,0\,1$, $1\,1\,0$ respectively; P the pole $h\,k\,l$.

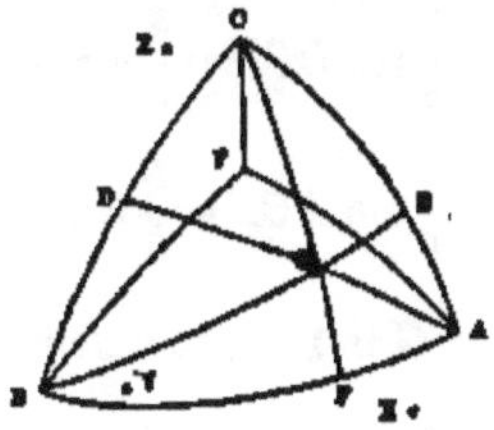

Since X, Y, Z are the poles of the great circles BC, CA, AB, the arcs XP, YP, ZP are the complements of the perpendiculars from P on BC, CA, AB. Therefore

$$\cos XP = \sin CP \sin BCP = \sin BP \sin CBP,$$

$$\cos YP = \sin AP \sin CAP = \sin CP \sin ACP,$$

$$\cos ZP = \sin BP \sin ABP = \sin AP \sin BAP.$$

But $\quad \frac{a}{h}\cos XP = \frac{b}{k}\cos YP = \frac{c}{l}\cos ZP.$ Therefore

$$\frac{a}{h}\sin CP \sin BCP = \frac{a}{h}\sin BP \sin CBP$$

$$= \frac{b}{k}\sin AP \sin CAP = \frac{b}{k}\sin CP \sin ACP$$

$$= \frac{c}{l}\sin BP \sin ABP = \frac{c}{l}\sin AP \sin BAP.$$

Hence

$$\frac{k}{b}\sin BAP = \frac{l}{c}\sin CAP,$$

$$\frac{l}{c}\sin CBP = \frac{h}{a}\sin ABP,$$

$$\frac{h}{a}\sin ACP = \frac{k}{b}\sin BCP.$$

The symbol of G is 1 1 1, therefore

$$\frac{1}{b}\sin BAG = \frac{1}{c}\sin CAG,$$

$$\frac{1}{c}\sin CBG = \frac{1}{a}\sin ABG,$$

$$\frac{1}{a}\sin ACP = \frac{1}{b}\sin BCG.$$

But

$$\sin BD \sin BDG = \sin AB \sin BAG,$$
$$\sin CD \sin CDG = \sin CA \sin CAG,$$
$$\sin CE \sin CEG = \sin BC \sin CBG,$$
$$\sin AE \sin AEG = \sin AB \sin ABG,$$
$$\sin AF \sin AFG = \sin CA \sin ACG,$$
$$\sin BF \sin BFG = \sin BC \sin BCG,$$
$$\sin BDG = \sin CDG, \quad \sin CEG = \sin AEG,$$
$$\sin AFG = \sin BFG.$$

Therefore

$$\frac{a}{b}=\frac{\sin AB}{\sin CA}\frac{\sin CD}{\sin BD},\quad \frac{a}{c}=\frac{\sin BC}{\sin AB}\frac{\sin AE}{\sin CE},\quad \frac{b}{a}=\frac{\sin CA}{\sin BC}\frac{\sin BF}{\sin AF}.$$

Hence

$$\frac{\sin CAP}{\sin BAP}=\frac{k}{l}\frac{\sin AB}{\sin CA}\frac{\sin CD}{\sin BD},$$

$$\frac{\sin ABP}{\sin CBP}=\frac{l}{h}\frac{\sin BC}{\sin AB}\frac{\sin AE}{\sin CE},$$

$$\frac{\sin BCP}{\sin ACP}=\frac{h}{k}\frac{\sin CA}{\sin BC}\frac{\sin BF}{\sin AF}.$$

Therefore, putting

$$\tan\theta=\frac{k}{l}\frac{\sin AB}{\sin CA}\frac{\sin CD}{\sin BD},$$

$$\tan\phi=\frac{l}{h}\frac{\sin BC}{\sin AB}\frac{\sin AE}{\sin CE},$$

$$\tan\psi=\frac{h}{k}\frac{\sin CA}{\sin BC}\frac{\sin BF}{\sin AF},$$

we obtain

$$\tan(BAP-\tfrac{1}{2}BAC)=\tan\tfrac{1}{2}BAC\tan(\tfrac{1}{4}\pi-\theta),$$

$$\tan(CBP-\tfrac{1}{2}CBA)=\tan\tfrac{1}{2}CBA\tan(\tfrac{1}{4}\pi-\phi),$$

$$\tan(ACP-\tfrac{1}{2}ACB)=\tan\tfrac{1}{2}ACB\tan(\tfrac{1}{4}\pi-\psi).$$

By means of these equations we can find the angles which the arcs *AP*, *BP*, *CP* make with the adjacent sides of the triangle *ABC*, and then, by the rules of spherical trigonometry, the arcs *AP*, *BP*, *CP* which determine the position of the pole *P*.

176. Multiplying together the expression for the ratios of the parameters in terms of the sides of the triangle *ABC*, and their segments (175), we obtain

$$\sin BD\sin CE\sin AF=\sin CD\sin AE\sin BF.$$

If we suppose five of the six arcs BD, CD, CE, AE, AF, BF to be known, the remaining arc will be given by this equation. The sides of the triangle ABC, and, therefore, its angles also are known. Therefore the angles which the axes make with one another, being the supplements of the angles of the triangle ABC, are known, and the ratios of the parameters are given in terms of the sides of ABC and their segments. Hence, any five of the six arcs BD, CD, CE, AF, BF may be taken for the elements of the crystal.

177. The six segments may also be deduced from one of the sides of ABC, and the segments of the other two sides. Suppose BC and the segments of CA, AB given. Then

$$\frac{\sin BD}{\sin CD} = \frac{\sin AE}{\sin CE}\,\frac{\sin BF}{\sin AF}.$$ Therefore, putting

$$\tan\theta = \frac{\sin AE}{\sin CE}\,\frac{\sin BF}{\sin AF},$$

we have $$\tan(CD - \tfrac{1}{2}BC) = \tan\tfrac{1}{2}BC\tan(\tfrac{1}{4}\pi - \theta).$$

Whence CD and BD are known.

178. The place of a pole in one of the zone-circles BC, CA, AB, or in any zone-circle containing three poles joined by arcs of known length, may be found by (13) or (14). In this manner it is usually possible to determine the places of all the poles of a crystal belonging to the anorthic or any other system.

179. Let L, M, N, O be any four poles of which no three are in one zone-circle; e f g, h k l, p q r the symbols of the zone-circles MN, NL, LM respectively; $m\,n\,o$ the symbol of O; $u\,v\,w$ the symbol of P. Suppose five of the six arcs joining every two of the poles L, M, N, O to be given. The remaining arc and the angles MLN, MLO, LMN, LMO can be found by the methods of spherical trigonometry. Then (18),

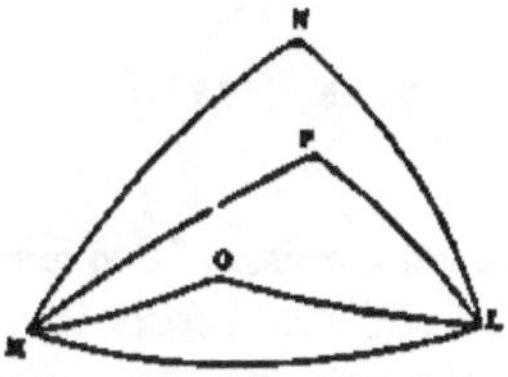

putting $\tan\theta = \dfrac{pm+qn+ro}{pu+qv+rw}\,\dfrac{hu+kv+lw}{hm+kn+lo}\,\dfrac{\sin(MLN-MLO)}{\sin MLO}$,

and $\tan\phi = \dfrac{pm+qn+ro}{pu+qv+rw}\,\dfrac{eu+fv+gw}{em+fn+go}\,\dfrac{\sin(LMN-LMO)}{\sin LMO}$,

we have $\tan(MLP-\tfrac{1}{2}MLN) = \tan\tfrac{1}{2}MLN\tan(\tfrac{1}{4}\pi-\theta)$,

and $\tan(LMP-\tfrac{1}{2}LMN) = \tan\tfrac{1}{2}LMN\tan(\tfrac{1}{4}\pi-\phi)$.

Hence, knowing *LM*, and the angles *MLP*, *LMP*, we can find the arcs *LP*, *MP* which determine the position of *P*.

180. When the position of any pole *P* is given with respect to any two of four given poles, no three of which are in one zone-circle, the ratios of the indices of *P* are given by the equations in (175) or (179).

181. Let *L*, *M* either have the same signification as in (179), or be any two of the poles *A*, *B*, *C* in (175); *P*, *Q* any two poles, the symbols of which are given. Let the angles *MLP*, *MLQ* and the arcs *LP*, *LQ* be found by (175) or (179). Then, knowing the sides *LP*, *LQ*, and the included angle *PLQ*, the third side *PQ* may be found.

182. When five of the six arcs joining every two of the poles *L*, *M*, *N*, *O* are given, the arc joining any two poles may be found by (179) and (181). Hence we can find the arcs *BD*, *CD*, *CE*, *AE*, *AF*, *DF*, or the angular elements of the crystal.

CHAPTER VIII.

TWIN CRYSTALS.

183. A TWIN crystal consists of two crystals joined together in such a manner, that one would come into the position of the other, by revolving through two right angles round an axis which is either normal to a possible face, or parallel to the axis of a possible zone, of each of the two crystals. This axis is called the twin axis. When it is normal to a possible face, the face is called a twin face. It frequently happens that, in twin crystals of any system except the anorthic, the twin axis is normal to a possible face, and also parallel to the axis of a possible zone, of each of the two crystals.

184. Let *T*, *T'* be a diameter of the sphere of projection parallel to the twin axis; *P*, *p* any corresponding poles of the

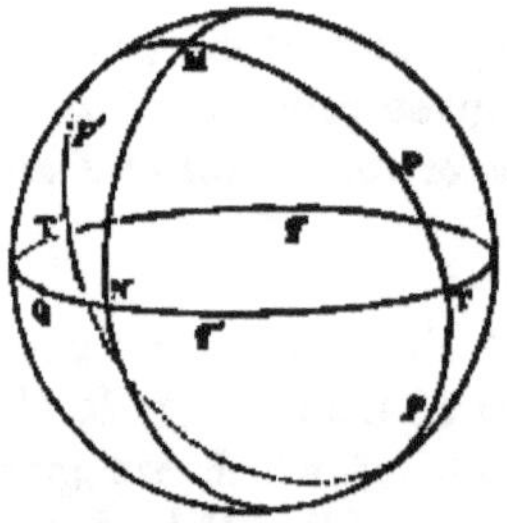

two crystals. Since *p* may be made to coincide with *P* by turning the crystal to which *p* belongs through two right angles round TT', the arc Tp = arc TP, and the angle $PTp = 180°$, or Pp is an arc of a great circle bisected in T. In like manner Q, q being any other corresponding poles of the two crystals, the arc Qq will be bisected in T. If p', q' be the poles opposite to p, q respectively, it is manifest that Pp', Qq' are bisected at right angles by the great circle MN having T, T' for its poles. Hence the opposite poles of the two crystals are symmetrically arranged with respect to a great circle having its poles in the twin axis.

185. In order to find the twin axis in any given twin crystal, when it cannot be found by simple inspection, we must determine by measurement or by the observation of zones, the intersections of two great circles each of which passes through corresponding or opposite poles of the two crystals. If the diameter of the sphere joining the intersections of the two circles be normal to corresponding faces or be the axis of corresponding zones of the two crystals, it will be the twin axis.

Let P, Q be any two poles of one crystal; p, q the corresponding poles of the other; p', q' the poles opposite to p, q; T, T' the intersections of the great circles pP, qQ. Then, if TT' be normal to a possible face or parallel to the axis of a possible zone of each of the two crystals, it will be the twin axis.

186. When the twin axis, and the angles of one of the crystals are given, the arc joining any pole of one crystal, and any pole of the other, can be readily determined. First let P, p be corresponding poles of the two crystals, p' the pole opposite to p. Then $Tp = TP$, and pPp' is a semicircle, therefore $Pp = 2TP$, and $Pp' = 180° - 2TP$. When TP is greater than a quadrant, Pp' is negative, and the faces P, p' will form a re-entrant angle. Next let P, Q be any two poles of one crystal; p, q the corresponding poles of the other. From the given arcs TP, TQ, PQ the angle PTQ is known, and $pTQ = 180° - PTQ$. Therefore, knowing TQ, Tp and the angle pTQ, the arc pQ may be found.

CHAPTER IX.

GEOMETRICAL INVESTIGATION OF THE PROPERTIES OF A SYSTEM OF PLANES.

187. Let any three straight lines in one plane, intersecting one another in the points A, B, C, meet any other straight line in the same plane, in D, E, F, the points D, E, F being in the

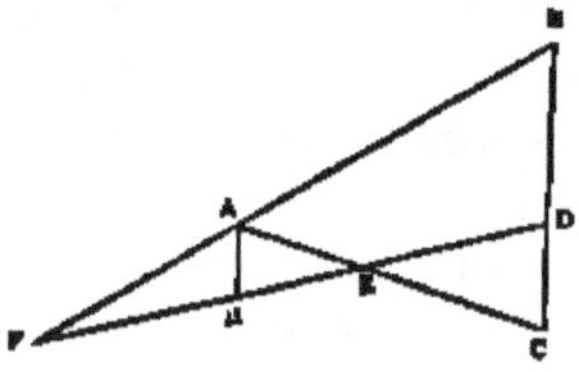

lines respectively opposite to A, B, C. From A draw AH parallel to BC, meeting DF in H. By similar triangles $AF : AH = BF : BD$, and $AH : AE = CD : CE$. Hence

$$CD \, . \, AE \, . \, BF = BD \, . \, CE \, . \, AF.$$

188. Let OX, OY, OZ be any three straight lines passing through a given point O, and not all in one plane; a, b, c any three straight lines given in magnitude; h, k, l any three integers, positive or negative or zero, one at least being finite. Let the symbol $h \; k \; l$ be used to denote the plane HKL which meets OX, OY, OZ in the points H, K, L such that

$$h\frac{OH}{a} = k\frac{OK}{b} = l\frac{OL}{c},$$

OH, OK, OL being measured along OX, OY, OZ, or in the opposite directions, according as the corresponding numbers h, k, l are positive or negative. And suppose a system of such planes to be obtained by giving h, k, l different numerical values. Let the point O be called the *origin* of the system of planes; OX, OY, OZ its *axes*; a, b, c, or any three straight lines in the same ratio, its *parameters*; h, k, l, or any three integers in the same ratio, and with the same signs, the *indices* of the plane HKL. When an index is taken negatively, the negative sign will be placed over the index usually, but not invariably. It is evident that when one of the indices of a plane becomes 0, the point of intersection of the plane with the corresponding axis will be indefinitely distant from the origin, and the plane will be parallel to that axis; also, that when two

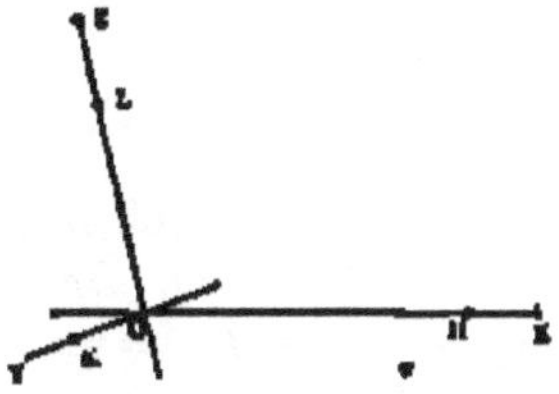

of the indices become 0, the plane will be parallel to the plane containing the two corresponding axes. The planes $h\,k\,l$, $\bar{h}\,\bar{k}\,\bar{l}$ are obviously parallel, and on opposite sides of the origin. Either symbol may be used to denote a plane through O, parallel to the plane HKL. The straight line in which any two planes intersect will be called an *edge*.

189. Let O be the origin of a system of planes; OX, OY, OZ its axes; a, b, c its parameters. Let $OB = b$; and let the planes $h\,k\,l$, $p\,q\,r$, passing through B, intersect one another in

the edge BM meeting the plane ZOX in M; and let them meet OZ in L, R, and OX in H, P. Then (188)

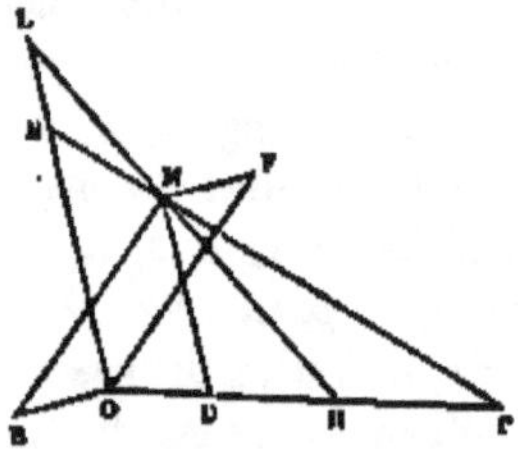

$$\frac{h}{a} OH = \frac{k}{b} OB = \frac{l}{c} OL, \text{ and } \frac{p}{a} OP = \frac{q}{b} OB = \frac{r}{c} OR.$$

Therefore $l.OL = kc$, $h.OH = ka$, $r.OR = qc$, $p.OP = qa$. Hence, $lr.LR = (kr - lq)\,c$, $hp.HP = (hq - kp)\,a$. But (187) $HM.OP.LR = HP.OR.LM$. Therefore, putting $\mathrm{u} = kr - lq$, $\mathrm{v} = lp - hr$, $\mathrm{w} = hq - kp$, we have

$$\mathrm{w}l.LM = \mathrm{u}h.HM, \quad \mathrm{w}l.LH = -\mathrm{v}k.HM, \quad \mathrm{u}h.LH = -\mathrm{v}k.LM.$$

Draw MD parallel to OZ, meeting OX in D. By similar triangles $OD : LM = OH : LH$, and $DM : HM = OL : LH$. Hence $-\mathrm{v}.OD = \mathrm{u}a$, and $-\mathrm{v}.DM = \mathrm{w}c$. Draw MF equal and parallel to OB, on the opposite side of the plane LOH. Then $-\mathrm{v}.MF = \mathrm{v}.OB = \mathrm{v}b$. The edge BM is obviously parallel to OF, the diagonal of a parallelopiped, the edges of which are respectively coincident with the axes OX, OY, OZ, and equal to OD, MF, DM, and therefore proportional to $-\mathrm{v}.OD$, $-\mathrm{v}.MF$, $-\mathrm{v}.DM$, or to $\mathrm{u}a$, $\mathrm{v}b$, $\mathrm{w}c$.

The edge BM, and any straight line parallel to BM, will be denoted by the symbol u v w, or by any whole numbers in the same ratio. The integers u, v, w, or any other integers in the same ratio, will be called the indices of the edge BM, or of any straight line parallel to BM.

190. Since a plane of the system may be parallel to any given edge, and also to any one of the other edges of the system, it follows that a number of planes may exist parallel to a given edge, and, therefore, intersecting one another in parallel lines. Such an assemblage of planes is called a *zone*. A straight line through the origin parallel to the edge in which any two of its planes intersect one another, is called the *axis* of the zone. A zone and its axis will be denoted by the symbol of the edge in which any two of its planes intersect. Hence (189), $h\,k\,l$, $p\,q\,r$ being the symbols of any two planes of the zone, not parallel to one another, the symbol of the zone will be u v w, where

$$\mathrm{u} = kr - lq, \quad \mathrm{v} = lp - hr, \quad \mathrm{w} = hq - kp.$$

It appears from (188) that the symbols of the planes YOZ, ZOX, XOY are 1 0 0, 0 1 0, 0 0 1 respectively. Hence, the symbol of OX, the intersection of the planes 0 1 0, 0 0 1, will be 1 0 0; the symbol of OY, the intersection of the planes 0 0 1, 1 0 0, will be 0 1 0; and the symbol of OZ, the intersection of the planes 1 0 0, 0 1 0, will be 0 0 1.

191. Let the plane $u\,v\,w$, meeting the axes of the system of planes in U, V, W, be parallel to the edge p q r. If VM be drawn parallel to the edge p q r, it will lie in the plane UVW, and its symbol will be p q r. Let VM meet WU in M. Then (189) $\mathrm{p}u \,.\, WU + \mathrm{q}v \,.\, WM = 0$, and $\mathrm{r}w \,.\, WU + \mathrm{q}v \,.\, UM = 0$; whence, adding, and observing that $WM + UM = WU$, we obtain

$$\mathrm{p}u + \mathrm{q}v + \mathrm{r}w = 0.$$

This equation expresses the condition which must be satisfied in order that the plane $u\,v\,w$ may belong to the zone p q r. Any three integers either positive or negative or zero, one at least being finite, which satisfy the preceding equation when substituted for u, v, w, are the indices of a plane in the zone p q r; and any three such integers which satisfy the same equation when substituted for p, q, r, are the indices of a zone containing the plane $u\,v\,w$.

192. Let h k l, p q r be the symbols of any two edges. In OY take $OV=b$, and through V draw VM, VS parallel to the edges h k l, p q r respectively, meeting the plane ZOX in M, S. Let MS meet OZ, OX in W, U. Draw MD, SG parallel to OZ, meeting OX in D, G. The symbols of VM, VS are h k l, p q r, therefore (189)

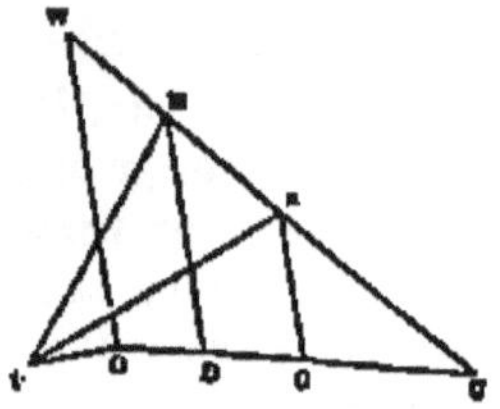

$$\mathrm{k}.OD=-\mathrm{h}a,\quad \mathrm{k}.DM=-\mathrm{l}c,\quad \mathrm{q}.OG=-\mathrm{p}a,\quad \mathrm{q}.GS=-\mathrm{r}c.$$

By similar triangles

$$OW : OU = DM : DU = DM - GS : OG - OD.$$

Hence $\mathrm{k}\,(\mathrm{kr}-\mathrm{lq}).DU=-\mathrm{l}\,(\mathrm{hq}-\mathrm{kp})\,a$; also, observing that $OU=OD+DU$, we obtain $(\mathrm{kr}-\mathrm{lq}).OU=(\mathrm{lp}-\mathrm{hr})\,a$, and $(\mathrm{hq}-\mathrm{kp}).OW=(\mathrm{lp}-\mathrm{hr})\,c$. Therefore

$$\frac{u}{a}\,OU=\frac{v}{b}\,OV=\frac{w}{c}\,OW,$$

where $u=\mathrm{kr}-\mathrm{lq}$, $v=\mathrm{lp}-\mathrm{hr}$, $w=\mathrm{hq}-\mathrm{kp}$.

Since u, v, w are integers, the plane UVW which is parallel to the edges h k l, p q r, is a plane of the system.

193. Let the plane $u\,v\,w$ meet the axes of the system in U, V, W, and the zone-axis e f g in P. Draw WP meeting UV in N, UP meeting VW in L, and PQ parallel to OU, meeting the plane VOW in Q. The symbols of OW, OU, OP are 0 0 1, 1 0 0, e f g respectively. Therefore the symbol of the plane WOP will be f $\bar{\mathrm{e}}$ 0, and that of the plane UOP will be 0 $\bar{\mathrm{g}}$ f. The symbol of the plane UVW is $u\,v\,w$. Hence, the symbol of the edge WN will be $-\mathrm{e}w$, $-\mathrm{f}w$, $\mathrm{e}u+\mathrm{f}v$, and the symbol of the edge UL will be $\mathrm{f}v+\mathrm{g}w$, $-\mathrm{f}u$, $-\mathrm{g}u$. The edges WN, UL are in the plane UVW, therefore (189) $\mathrm{e}u.UN=\mathrm{f}v.VN$, and $\mathrm{f}v.VW=(\mathrm{f}v+\mathrm{g}w).WL$. But by (187) $UP.WL.VN=PL.VW.UN$. Therefore $\mathrm{e}u.UP=(\mathrm{f}v+\mathrm{g}w).PL$. There-

fore $eu . UL = (eu + fv + gw) . PL$. But $QP : OU = PL : UL$. Therefore $eu . OU = (eu + fv + gw) . QP$. In like manner, if

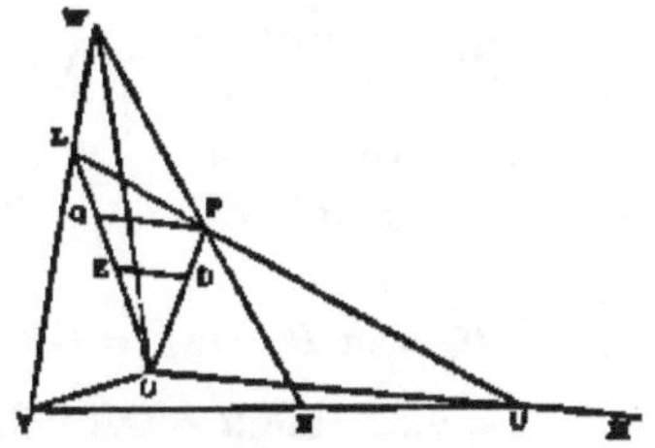

the plane hkl meet OU in H, and OP in D, and if DE be drawn parallel to OU, meeting the plane VOW in E, we shall have $eh . OH = (eh + fk + gl) . ED$. But $OP : OD = QP : ED$. Therefore $(eu + fv + gw) . OP : (eh + fk + gl) . OD = u . OU : h . OH$. Hence, if the zone-axis p q r meet the planes $u\,v\,w$, $h\,k\,l$ in R, F, we shall have

$$(pu + qv + rw) . OR : (ph + qk + rl) . OF = u . OU : h . OH.$$

Therefore $$\frac{eu + fv + gw}{pu + qv + rw} \frac{OP}{OR} = \frac{eh + fk + gl}{ph + qk + rl} \frac{OD}{OF}.$$

The preceding equation will still be true, if we suppose OP, OR to be the edges e f g, p q r passing through any point O which is not the origin of the system of planes. For OP, OR will be parallel to the zone-axes e f g, p q r respectively, and, therefore, the ratios $OP : OR$, $OD : OF$ will be the same in either case.

194. If DF, PR intersect in K, we shall have

$$KP \sin P = KD \sin D, \quad KR \sin R = KF \sin F,$$

and $$OD \sin D = OF \sin F, \quad OP \sin P = OR \sin R.$$

Hence $$KP . KF : KD . KR = OP . OF : OD . OR.$$

Therefore (193), $$\frac{eu + fv + gw}{pu + qv + rw} \frac{KP}{KR} = \frac{eh + fk + gl}{ph + qk + rl} \frac{KD}{KF}.$$

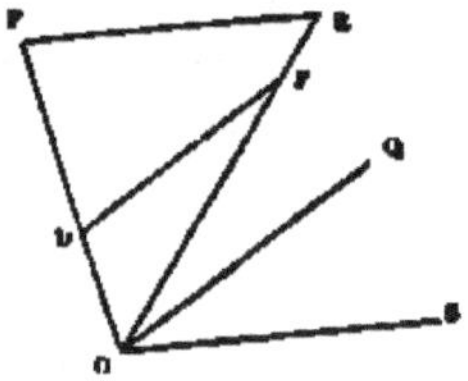

195. Let the planes $h\,k\,l$, $u\,v\,w$ meet the zone-axis e f g in D, P, and the zone-axis p q r in F, R, O being the origin. Draw OQ, OS parallel to DF, PR respectively. Then OQ, OS will be the axes of zones containing the planes $h\,k\,l$, $u\,v\,w$, and will be in the plane POR;

$$\sin POQ : \sin ROQ = \sin D : \sin F = OF : OD, \text{ and}$$

$$\sin POS : \sin ROS = \sin P : \sin R = OR : OP; \text{ also (193),}$$

$$\frac{eu+fv+gw}{pu+qv+rw}\,\frac{OP}{OR} = \frac{eh+fk+gl}{ph+qk+rl}\,\frac{OD}{OF}. \text{ Therefore}$$

$$\frac{\sin POQ}{\sin POS}\,\frac{\sin ROS}{\sin ROQ} = \frac{eh+fk+gl}{eu+fv+gw}\,\frac{pu+qv+rw}{ph+qk+rl},$$

where OP, OQ, OR, OS are four zone-axes in one plane; OP, OR the axes of the zones e f g, p q r; and OQ, OS the axes of zones containing the planes $h\,k\,l$, $u\,v\,w$.

It appears from (13) that the left-hand side of the preceding equation can be put under the form

$$(\cot POS - \cot POR) : (\cot POQ - \cot POR),$$

which is manifestly positive, except when one only of the zone-axes OP, OR lies between OQ and OS.

196. Let P, Q, R, S be four planes in one zone. Let a plane passing through the origin O, normal to the axis of the zone, meet the planes Q, S in df, pr; and planes passing through O, parallel to the planes P, R, in dp, fr. Let $h\,k\,l$, $u\,v\,w$ be the symbols of the planes Q, S; e f g, p q r the symbols of any zones containing the planes P, R respectively, except the zone containing P and R. Then the zone-axes e f g, p q r lie in the planes parallel to the planes P, R respectively; Od, Op are

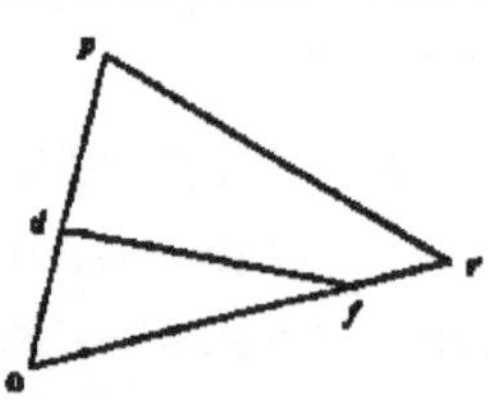

proportional to the portions of the zone-axis e f g intercepted between O and the planes Q, S; and Of, Or are proportional to the portions of the zone-axis p q r intercepted between O and the planes Q, S. Therefore (193),

$$\frac{eu + fv + gw}{pu + qv + rw}\,\frac{Op}{Or} = \frac{eh + fk + gl}{ph + qk + rl}\,\frac{Od}{Of}.$$

If PQ, PS, RQ, RS be taken to denote the angles which the planes Q, S make with the planes P, R, we shall have

$\sin PQ = \sin d$, $\sin PS = \sin p$, $\sin RQ = \sin f$, $\sin RS = \sin r$.

But $\sin p : \sin r = Or : Op$, and $\sin d : \sin f = Of : Od$. Hence

$$\frac{\sin PQ}{\sin PS}\,\frac{\sin RS}{\sin RQ} = \frac{eh + fk + gl}{eu + fv + gw}\,\frac{pu + qv + rw}{ph + qk + rl},$$

where P, Q, R, S are four planes in one zone; e f g, p q r the symbols of zones containing the planes P, R; and $h\,k\,l$, $u\,v\,w$ the symbols of the planes Q, S.

It may be shewn, as in (195), that the left-hand side of the preceding equation is positive, except when one only of the planes P, R lies between the planes Q, S.

197. Let e f g, h k l, p q r be the symbols of three zone-axes OP, OQ, OR meeting the plane $m\,n\,o$ in the points D, E, F, and the plane $u\,v\,w$ in the points P, Q, R. Then (193),

$$\frac{eu + fv + gw}{em + fn + go}\,\frac{OP}{OD} = \frac{hu + kv + lw}{hm + kn + lo}\,\frac{OQ}{OE} = \frac{pu + qv + rw}{pm + qn + ro}\,\frac{OR}{OF}.$$

But if $m'n'o'$, $u'v'w'$ be the symbols of the planes $m\,n\,o$, $u\,v\,w$, when referred to the zone-axes e f g, h k l, p q r, as axes of the system of planes, we shall have

$$\frac{u'}{m'}\,\frac{OP}{OD} = \frac{v'}{n'}\,\frac{OQ}{OE} = \frac{w'}{o'}\,\frac{OR}{OF}.$$

Hence, comparing identical terms, two equations are obtained which are satisfied by making

$$m' = em + fn + go,\quad u' = eu + fv + gw,$$
$$n' = hm + kn + lo,\quad v' = hu + kv + lw,$$
$$o' = pm + qn + ro,\quad w' = pu + qv + rw.$$

199. Let m n o, u v w be the symbols of the zone-axes OG, OP. Through G draw the planes efg, hkl, pqr meeting OR in R, S, T respectively. Then (193),

$$\frac{ue+vf+wg}{me+nf+og}\,\frac{OR}{OG}=\frac{uh+vk+wl}{mh+nk+ol}\,\frac{OS}{OG}=\frac{up+vq+wr}{mp+nq+or}\,\frac{OT}{OG}.$$

Let m′n′o′, u′v′w′ be the symbols of OG, OR, when referred to axes parallel to the intersections of the planes efg, hkl, pqr. The symbols of these two planes when referred to the new axes will become 100, 010, 001 respectively. Therefore (193),

$$\frac{u'}{m'}\,\frac{OR}{OG}=\frac{v'}{n'}\,\frac{OS}{OG}=\frac{w'}{o'}\,\frac{OT}{OG}.$$

Hence, comparing identical terms, we obtain two equations which are satisfied by making

$$m'=em+fn+go,\quad u'=eu+fv+gw,$$
$$n'=hm+kn+lo,\quad v'=hu+kv+lw,$$
$$o'=pm+qn+ro,\quad w'=pu+qv+rw.$$

CHAPTER X.

ANALYTICAL INVESTIGATION OF THE PROPERTIES OF A SYSTEM OF PLANES.

199. As in (188), let OX, OY, OZ be any three axes not all in one plane; a, b, c any three straight lines given in magnitude; h, k, l any three integers, positive or negative or zero, one of them at least remaining finite; H, K, L three points in OX, OY, OZ respectively, subject to the condition

$$h\frac{OH}{a} = k\frac{OK}{b} = l\frac{OL}{c}.$$

Then, d being any positive quantity, the equation to the plane HKL will be

$$h\frac{x}{a} + k\frac{y}{b} + l\frac{z}{c} = d.$$

Let the plane HKL be denoted by the symbol $h\,k\,l$, or by any three integers respectively proportional to h, k, l, and having the same sign, the numbers h, k, l being called the indices of the plane HKL. A system of planes being formed by giving h, k, l different numerical values, let the straight lines a, b, c be called the parameters of the system of planes.

200. The equations to the planes $h\,k\,l$, $p\,q\,r$ are

$$h\frac{x}{a} + k\frac{y}{b} + l\frac{z}{c} = d, \quad p\frac{x}{a} + q\frac{y}{b} + r\frac{z}{c} = t,$$

where d, e are positive quantities. The intersection of the planes $h\,k\,l$, $p\,q\,r$ will, therefore, be parallel to the line which has for its equations

$$\frac{x}{\mathrm{u}a} = \frac{y}{\mathrm{v}b} = \frac{z}{\mathrm{w}c},$$

where $\quad \mathrm{u} = kr - lq, \quad \mathrm{v} = lp - hr, \quad \mathrm{w} = hq - kp.$

This straight line, or any straight line parallel to it, will be denoted by the symbol u v w, or by any three integers proportional to u, v, w. These three numbers will be called the indices of the line.

This straight line is obviously the diagonal OK of a parallelopiped having its edges OU, OV, OW coincident with the axes, and equal to ua, vb, wc respectively.

201. Any number of planes intersecting one another in parallel lines are said to constitute a zone. A straight line through the origin, parallel to the intersection of any two planes of a zone, and, therefore, parallel to each of the planes of the zone, will be called the axis of the zone. A zone, and its axis, will be denoted by the symbol of a line parallel to the intersection of any two planes of the zone. Hence (200) the symbol of the zone containing the planes $h\,k\,l$, $p\,q\,r$ will be u v w, where $\mathrm{u} = kr - lq, \quad \mathrm{v} = lp - hr, \quad \mathrm{w} = hq - kp.$

202. Let the zone-axis p q r be parallel to the plane u v w. The equations to the zone-axis and plane are

$$\frac{x}{\mathrm{p}a} = \frac{y}{\mathrm{q}b} = \frac{z}{\mathrm{r}c}, \quad \text{and} \quad \mathrm{u}\frac{x}{a} + \mathrm{v}\frac{y}{b} + \mathrm{w}\frac{z}{c} = d;$$

and the zone-axis is parallel to the plane. Hence

$$\mathrm{p}u + \mathrm{q}v + \mathrm{r}w = 0.$$

Any three positive or negative integers, including one or two zeros, which satisfy the preceding equation, when substituted for u, v, w, are the indices of a plane in the zone p q r; and any three such integers which satisfy the same equation, when substituted for p, q, r, are the indices of a zone containing the plane $u\,v\,w$.

203. The equations to the zone-axes h k l, p q r are

$$\frac{x}{ha}=\frac{y}{kb}=\frac{z}{lc}, \text{ and } \frac{x}{pa}=\frac{y}{qb}=\frac{z}{rc}.$$

Hence, if a plane be drawn parallel to the zone-axes h k l, p q r, its equation will be

$$u\frac{x}{a}+v\frac{y}{b}+w\frac{z}{c}=d,$$

where $\quad u=kr-lq, \quad v=lp-hr, \quad w=hq-kp.$

Therefore, since u, v, w are integers, a plane parallel to any two zone-axes will be a plane of the system.

204. Let e f g, p q r be the symbols of the zone-axes *OP*, *OR* meeting the plane *h k l* in *D*, *F*, and the plane *u v w* in *P*, *R*. Let planes be drawn parallel to *YOZ*, through the points *D*, *P*, *F*, *R*, meeting *OX* in the points *D′*, *P′*, *F′*, *R′*. The equations to the zone-axes e f g, p q r are

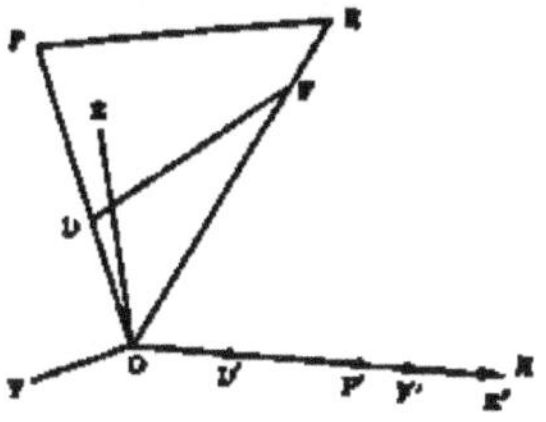

$$\frac{x}{ea}=\frac{y}{fb}=\frac{z}{gc}, \text{ and } \frac{x}{pa}=\frac{y}{qb}=\frac{z}{rc};$$

and the equations to the planes *h k l*, *u v w* are

$$h\frac{x}{a}+k\frac{y}{b}+l\frac{z}{c}=d, \text{ and } u\frac{x}{a}+v\frac{y}{b}+w\frac{z}{c}=t.$$

The distances OD', OP', OF', OR' are the values of x at the points in which the zone-axes e f g, p q r intersect the planes $h\,k\,l$, $u\,v\,w$. Therefore

$$(eh + fk + gl)\,.\,OD' = ead, \quad (eu + fv + gw)\,.\,OP' = eat,$$

$$(ph + qk + rl)\,.\,OF' = pad, \quad (pu + qv + rw)\,.\,OR' = pat.$$

And by similar triangles

$OD' : OD = OP' : OP$, and $OF' : OF = OR' : OR$. Therefore

$$\frac{eu + fv + gw}{pu + qv + rw}\,\frac{OP}{OR} = \frac{eh + fk + gl}{ph + qk + rl}\,\frac{OD}{OF}.$$

205. From the preceding equation the expressions for the anharmonic ratios of four zone-axes in one plane, and of four planes in one zone, and the indices of a plane or a zone when the axes are changed, can be found as in (195), (196), (197) and (198).

www.ingramcontent.com/pod-product-compliance
Lightning Source LLC
Chambersburg PA
CBHW021948190326
41519CB00009B/1188
* 9 7 8 3 7 4 1 1 6 3 7 2 2 *